THE ONTOLOGICAL TURN

A recent and often controversial theoretical orientation that resonates strongly with wider developments in contemporary philosophy and social theory, the so-called ontological turn is receiving a great deal of attention in anthropology and cognate disciplines at present. This book provides the first anthropological exposition of this recent intellectual development. It traces the roots of the ontological turn in the history of anthropology and elucidates its emergence as a distinct theoretical orientation over the past few decades, showing how it has emerged in the work of Roy Wagner, Marilyn Strathern and Viveiros de Castro, as well as a number of younger scholars. Distinguishing this trajectory of thinking from related attempts to put questions of ontology at the heart of anthropological research, the book articulates critically the key methodological and theoretical tenets of the ontological turn, its prime epistemological and political implications, and locates it on the broader intellectual landscape of contemporary social theory.

MARTIN HOLBRAAD is Professor of Social Anthropology at University College London (UCL). He is the author of *Truth in Motion: The Recursive Anthropology of Cuban Divination* (2012) and co-editor of *Thinking Through Things: Theorizing Artefacts Ethnographically* (2007). Having studied the relationship between religious and political practices in Cuba since the late 1990s, he currently holds a European Research Council Consolidator Grant for a five-year project titled *Comparative Anthropologies of Revolutionary Politics*, leading a team of researchers to chart comparatively the formation of revolutionary personhood in selected countries of Latin America and the Middle East and North Africa.

MORTEN AXEL PEDERSEN is Professor of Social Anthropology at the University of Copenhagen. He is the author of *Not Quite Shamans: Spirit Worlds and Political Lives in Northern Mongolia* (2011), which received honourable mention for the Bateson Prize, and (with L. Højer) *Urban Hunters: Dreaming and Dealing in Times of Transition* (in press). From 2011–2016 he held a Sapere Aude Research Leader Grant from the Danish Research Council, sparking off his recent reseach on Lutheran Christian movements and vernacular political theology in Denmark.

NEW DEPARTURES IN ANTHROPOLOGY

New Departures in Anthropology is a book series that focuses on emerging themes in social and cultural anthropology. With original perspectives and syntheses, authors introduce new areas of inquiry in anthropology, explore developments that cross disciplinary boundaries, and weight in on current debates. Every book illustrates theoretical issues with ethnographic material drawn from current research or classic studies, as well as from literature, memoirs and other genres of reportage. The aim of the series is to produce books that are accessible enough to be used by college students and instructors, but will also stimulate, provoke and inform anthropologists at all stages of their careers. Written clearly and concisely, books in the series are designed equally for advanced students and a broader range of readers, inside and outside academic anthropology, who want to be brought up to date on the most exciting developments in the discipline.

The Ontological Turn

An Anthropological Exposition

MARTIN HOLBRAAD
University College London

MORTEN AXEL PEDERSEN
University of Copenhagen

CAMBRIDGE
UNIVERSITY PRESS

University Printing House, Cambridge CB2 8BS, United Kingdom

One Liberty Plaza, 20th Floor, New York, NY 10006, USA

477 Williamstown Road, Port Melbourne, VIC 3207, Australia

4843/24, 2nd Floor, Ansari Road, Daryaganj, Delhi – 110002, India

79 Anson Road, #06-04/06, Singapore 079906

Cambridge University Press is part of the University of Cambridge.

It furthers the University's mission by disseminating knowledge in the pursuit of education, learning and research at the highest international levels of excellence.

www.cambridge.org
Information on this title: www.cambridge.org/9781107503946
10.1017/9781316218907

First published 2017

Printed in the United States of America by Sheridan Books, Inc.

A catalogue record for this publication is available from the British Library.

ISBN 978-1-107-10388-7 Hardback
ISBN 978-1-107-50394-6 Paperback

Contents

v

Contents

Figures and Boxes

Figures

Boxes

Preface and Acknowledgements

A controversial theoretical and methodological approach that resonates with wider developments in contemporary philosophy and social theory, the so-called ontological turn has been the subject of heated debates in anthropology and cognate disciplines such as archaeology and Science and Technology Studies over recent years. Drawing together and taking stock of these debates, this book traces the origins of the ontological turn in the history of anthropology and elucidates its emergence as a distinct analytical method since the postmodern crises of the 1980s, articulating its core theoretical tenets as well as its methodological, ethical and political implications. Placing the ontological turn within the broader intellectual landscape of both past and present anthropological theorizing, the book addresses the following basic questions: What are the key methodological and theoretical tenets of the ontological turn? What critiques has it elicited, and what are the possible responses to them? What are its wider epistemological, political and ethical ramifications?

This book's central contention is that the ontological turn in anthropology must be understood as a strictly methodological proposal – that is, a technology of ethnographic description. As such, the ontological turn asks ontological questions without taking ontology (or indeed ontologies) as an answer. Far from prescribing and thus curtailing the horizon of anthropological inquiry in the name of an ultimate reality or essence that may ground it (i.e. providing an 'ontology' in the substantive sense), the ontological turn is the methodological injunction to keep this horizon perpetually open. Above all, it is the injunction to keep constitutively open the question of what any given object of ethnographic investigation might *be* and, therefore, how existing concepts and theories have to be modulated in order the better to articulate it. What *are* the objects and manners of anthropological inquiry, and what could they *become*, are the abidingly ontological questions that lend the 'turn' its name. The ontological turn is not concerned with what the 'really real' nature of the world is or similar orthodox philosophical or metaphysical agendas often associated with the word 'ontology'. Rather, the ontological turn poses ontological questions to solve epistemological problems. Only, as we shall see, it so happens that epistemology in anthropology has to be about ontology, too.

So, anthropology has always engaged with ontological questions, even if this has not always been clear to the authors of ethnographic texts or their readers. Indeed, another core claim of this book is that rather than a revolutionary rupture from the anthropological past, the turn to ontology with which its chapters are concerned involves releasing in their fullest form analytical potentials that have always been at the heart of the discipline's project, and which can be recognized in some of the greatest exponents of the distinct mode of thinking we call anthropological, including, say, Mauss, Evans-Pritchard, Lévi-Strauss and Schneider. More precisely, the ontological turn involves three analytical practices that have been characteristic of the anthropological project possibly since its inception, namely reflexivity, conceptualization and experimentation, each of which can be recognized in theoretical developments within, and engagements between, the discipline's three so-called great national traditions, namely, the American, the British and the French. While a thorough account of this trans-Atlantic traffic in anthropological ideas and perspectives will have to be provided elsewhere, this book seeks to trace the core theoretical developments and genealogies that eventually congealed into the ontological turn, represented in the work, respectively, of Roy Wagner, Marilyn Strathern and Eduardo Viveiros de Castro.

Setting forth these intellectual developments systematically and in clear language, and scrutinizing their basic theoretical and methodological assumptions, the ambition of the book is to provide a general introduction to a body of literature that is often regarded as esoteric and difficult to read, contributing also to setting the agenda for its potential future development. The hope is that such a discussion of the ontological turn's place in the broader intellectual landscape might help to move the debate about it away from the divisive and earth-scorching manner so characteristic of 'first generation' discussions about ontology within anthropology, including some of our own writings. Far from stoking the fire by putting forward another debating piece written in the rhetorical and provocative style characteristic of hot academic controversy, the ambition is to engage with the critics of the ontological turn by clarifying potential misunderstandings and making explicit assumptions that have hitherto remained largely tacit. Certainly, there is need for a thorough and, ideally, straightforward exposition of what this theoretical orientation is all about, conveying its core tenets and surveying its analytical possibilities as well as potential pitfalls. It is up to the reader to decide whether we have gone some way towards meeting this goal.

The idea to write this book was first conceived over lunch conversations with Matei Candea, Eduardo Kohn and Patrice Maniglier at the Comparative Metaphysics Colloquium at Cerisy, Normandy, in August 2013 (see Charbonnier, Salmon & Skafish 2016). We thank them and other scholars participating in this seminal event, including its three organizers Pierre Charbonnier, Gildas Salmon and Peter Skafish, as well as Philippe Descola, for inspiration and encouragement. For their support we would also like to thank the editors of this book series, Michael Lambek and Jonathan Spencer, as well as our Cambridge University Press editor Andrew Winnard and other people from the Press, including Bethany Gaunt and Mary Catherine Bongiovi. Thanks also to Flora

Botelho and Neil Wells for their help with preparing the manuscript and the index for publication.

For reading and commenting on drafts of our chapters we are enormously grateful to Benjamin Alberti, Kristoffer Albris, Mikkel Bille, Tom Boellstorff, Matthew Carey, Igor Cherstich, Jo Cook, David Cooper, Iracema Dulley, Alice Elliot, Astrid Grue, Agnieszka Halemba, Casper Bruun Jensen, Stine Krøijer, Chloe Nahum-Claudel, Morten Nielsen, Adam Reed, Joel Robbins, Julia Sauma, Mario Schmitt, Michael Scott, Charles Stewart, Soumhya Venkatesan and James Weiner, as well as members of the Cosmology, Religion, Ontology and Culture (CROC) research group at University College London (UCL), students in the Contemporary Anthropological Theory class at the University of Copenhagen and the Advanced Cultural Theory seminar at the University of California Santa Cruz and participants in seminars, workshops and conferences held in the United Kingdom, Denmark and other parts of Europe, the United States, Cuba, Mongolia and Japan where different versions of the arguments developed in this book have been presented.

In Chapters 2, 3, 5 and 6 we have drawn liberally on the following previously published works: pp 37–46 of Holbraad's monograph *Truth in Motion: The Recursive Anthropology of Cuban Divination* (Chicago: University of Chicago Press, 2012); sections of our co-written article "Planet M: the intense abstraction of Marilyn Strathern," published in two different versions, in *Cambridge Anthropology* Volume 28, Issue 3, pp. 43–65 (2009) and *Anthropological Theory* Volume 9, Issue 4, pp. 371–394 (2009); sections of Holbraad's article "Can the thing speak?," first published online on the Open Anthropology Cooperative Press (Working Papers Series #7, 2011), with further versions published in *Savage Objects*, edited by G. Pereira (Guimaraes: INCM, 2013), pp. 17–30, and *Objects and Materials: A Routledge Companion*, edited by P. Harvey *et al.* (London: Routledge, 2014), pp. 228–237; sections of Chapter 4 in Pedersen's monograph *Not Quite Shamans: Spirit Worlds and Political Lives in Northern Mongolia* (Ithaca: Cornell University Press, 2011);

sections of Pedersen's article "The task of anthropology is to invent relations" published in *Critique of Anthropology* Volume 32, Issue 1, pp. 59–65 (2012), and Holbraad's commentary "Response to Bruno Latour's 'Thou shall not freeze-frame,'" written in 2004 and available online at abaetenet. net/nansi. Where relevant, we thank the publishers of these works for permission to draw on them here.

Holbraad would like to record his gratitude to successive cohorts of students at UCL who interrogated early versions of many of the ideas developed in this book (this is 'anthropology B'!); to colleagues at the Instituto de Filosofia in Havana, as well as Leo, Nely and Maryanis for hosting him during fieldwork made possible by his ERC Consolidator grant, ERC-2013-CoG, 617970, CARP; and to Alice for her intellectual stimulation and all-round love and support throughout the period of the book's preparation. In addition to thanking Kimi, Sophie and Ines for their patience, support and love, Pedersen also wishes to thank his colleagues and especially his former and current doctoral students (Dan, Antonia, Christian, Stine, Ida, Lise, Sandra and My) at the Department of Anthropology in Copenhagen for comments, criticism, reading suggestions and stimulating ideas that have contributed to the formulation and refinement of this book's approach and argument, as well as the University of California Santa Cruz anthropology department for a visiting professorship and the Danish Research Council of the Humanities for a Sapere Aude Research Leader grant which made his 2014 visit to California financially possible and intellectually stimulating.

We thank all these people as well as all those whom we may have omitted here by mistake: this book truly could not have been written without them. We reserve special thanks to Marilyn Strathern, Eduardo Viveiros de Castro and Roy Wagner who generously read and commented on the chapters presenting their work. The standard proviso about all responsibility for errors of interpretation or other inaccuracies being entirely our own is particularly apt in this case.

Introduction: The Ontological Turn in Anthropology

Consider an anthropology student getting his head around Marcel Mauss's idea that Maori gifts are returned because they are taken to contain within them the spirit of the donor (Mauss 1990). Or E. E. Evans-Pritchard's suggestion that Zande oracles don't answer the question of how something happened, but rather of why it happened to a particular person at a particular time (Evans-Pritchard 1937). Or Clifford Geertz's notion that certain Balinese calendars do not measure quantitatively the distances between past, present and future, but rather render each day qualitatively different from the one before – a matter not of what day it is but of what kind of day it is (Geertz 1973).

Such emblematic arguments, we know, stand for particular traditions within the discipline (respectively, the French, British and American), and it is likely that the student will have been introduced to them in this way. Still, what the three examples have in common is that they all illustrate a manner of thinking that is quintessentially anthropological. Consider the initial impact these arguments may have on our student: to understand Maori gifts, Zande oracles or Balinese calendars, he now realizes, you must be prepared to question some of the most basic things you may have taken for granted. Suddenly, the distinction between people and things, the assumption that events are best explained by their causes, or the notion that time is something that passes, are all up for grabs. The 'a-ha!-moment' that each of these examples is meant to induce, then,

Introduction

is at once reflexive and profoundly relativizing: assumptions that may seem self-evident, even absolute, are compromised by exposure to ethnographic realities that challenge them.

Different traditions and theoretical approaches in anthropology do different things with this basic manner of anthropological thought. Some have used the a-ha!-moment to formulate theories of cultural relativism. Others have sought to defuse it by showing how initially surprising ideas and practices are understandable once we realize that they are local ways of doing things we all do: Ideas about the spirit of the gift really are a Maori way of thinking about the profit of exchange (Sahlins 1974); oracular pronouncements really are the Azande's way of apportioning blame for misfortunes (Evans-Pritchard 1937); and the 'non-linear time' of Balinese calendars is part of the ideological reproduction of local ruling elites (Bloch 1977). Often, this kind of no-nonsense pragmatism bleeds into more elaborate theoretical models, in terms of universal human traits or other underlying mechanisms that may explain cross-cultural variations – evolutionary exigencies, socio-political functions, deep symbolic structures, cognitive operations and the like. As a result, the student's moment of ethnographic insight is pressed into the service of a larger effort to understand how the human (social, cultural, etc.) world works – his a-ha!-moment of intellectual relativization traded in for the bigger eureka-moments of scientific discovery.

This book is about a strand of anthropological thinking that does something altogether different with the discipline's relativizing a-ha!-moments, namely to *run with them*. Instead of encasing them within generalizing theories about culture, society, human nature and so forth, or trying to explain them away with a good dose of common sense, this way of thinking in anthropology seeks deliberately to take these moments as far as they will go, making full virtue of their capacity to stop thinking in its tracks, unsettling what we think we know in favour of what we may not even have imagined. To take just Mauss's *The Gift* as an example, what happens if one takes a step further the suggestion that

Maori gifts cut against the common-sense distinction between people and things? Might one not try to be more precise than just pointing out that the distinction between people and things 'does not apply' in this case, or saying that here 'people and things are continuous with each other', or are 'part of each other' (what *is* it, after all, for a person and a thing to be 'continuous' or 'part of each other')? Rather, is what is needed here not a wholesale re-conceptualization of the very notions of 'people', 'things' and their 'relationships'? Indeed, considering that anthropology defines itself as the discipline that studies people *par excellence* (including their relationships with things), how far might these reconceptualizations modify the way we think about anthropology itself, as a discipline, in terms of its objects and scope, as well as its methods and its impact?

The present book is about the turn of anthropological thinking that such questions exemplify. With reference to recent debates about 'ontology' within anthropology and related disciplines, and with a desire to intervene in them, we call this 'the ontological turn' of anthropological thinking. Explaining why these terms – 'ontological' and 'turn' – are appropriate will be one of the tasks of the book. Indeed, the central idea that this book develops is that, taken as far as they will go, the a-ha!-moments of anthropology lead ultimately to ontological considerations – considerations, that is, with what the objects of anthropological inquiry, as well as the terms in which the inquiry is conducted, might be: what *is* a thing, what *is* a person, and what *is* their mutual relationship, are the inherently ontological questions that the ethnographic exposure to, say, Maori gift exchange precipitates. So, taken to their logical conclusion the relativizing effects of the a-ha!-moments of anthropology are ontological.

As we shall be demonstrating in our exposition of different contributions to this line of thinking, such moments of ontological relativization – moments in which one's assumptions about what any given object or term of inquiry might *be* are called into question – are *necessary* to anthropological analysis.

Introduction

To return to our example again, asking what people and things 'might be' in Maori gift exchange is to ask what they *must be* for these practices to make anthropological sense. It is to ask for, and generate, the conceptual and analytical apparatus that will permit us even to describe, let alone cogently comprehend, Maori gift exchange, or whatever other ethnographic materials are of concern to us. Without the conceptual agility that ontological relativization provides, we suggest, anthropology is resigned to misunderstanding, even misdescribing, the very ethnographic materials it seeks to elucidate.

So, this is the central concern of the ontological turn: It is about creating the conditions under which one can 'see' things in one's ethnographic material that one would not otherwise have been able to see. And that, we should emphasize from the start, is at its core a *methodological* intervention, as opposed to a metaphysical or indeed philosophical one.[1] In spite of its name, the ontological turn in anthropology is therefore decidedly *not* concerned with what the 'really real'

---

[1] To be sure, we shall be seeing at certain points in chapters to follow, the reflexive project of conceptualization on which this anthropological approach centres does draw some of its inspiration from philosophical ideas and proposals. And conversely, it is worth noting that the interest anthropologists of the ontological turn have shown in philosophy has been to a certain extent reciprocated. As Tanya Luhrmann has noted (2013), contemporary discussions about ontology in anthropology can be compared to notorious debates about rationality in the 1960s and 70s, in which a number of philosophers engaged in a lively dialogue with anthropologists in entertaining the possibility of alternative forms of reasoning of the kind Evans-Pritchard, most emblematically perhaps, had sought to articulate for Zande witchcraft (1937; e.g. see Winch 1967; Wilson 1974). While the rationality debate had a clear epicentre in Britain, recent philosophical interest in anthropologists' turn to ontology has come from more diverse sources, crossing even the proverbial divide between Analytical and Continental traditions (e.g. compare Paleček & Risjord 2013 and Sivado 2015 with Watson 2014, Surel 2014, Maniglier 2014, and Charbonnier et al. 2016). It should be noted that these debates have been conducted largely independently from the classic conversation between philosophers and social scientists about the ontology of social phenomena (e.g. Weber 1968; Durkheim 1982; Elster 1982), which in recent years has continued into philosophical and social theoretical discussions about 'social ontologies' (e.g. Searle 1995; 2006; Marcoulatos 2003; Friedman 2006; Fullbrook 2008; Lawson 2012).

nature of the world is or any similar metaphysical quest. Rather, it is a methodological project that poses ontological questions to solve epistemological problems. Only, as pointed out in the Preface, it so happens that epistemology in anthropology has to be about ontology, too.

In particular, the ontological turn is a response to that most fundamental anthropological question: How do I enable my ethnographic material to reveal itself to me by allowing it to dictate its own terms of engagement, so to speak, guiding or compelling me to see things that I had not expected, or imagined, to be there? Through what analytical techniques might such an ethnographic sensibility be cultivated? This, of course, is a version of anthropologists' most abiding methodological concern, namely with how to neutralize the danger of one's own presuppositions constraining or even predetermining one's capacity to describe, interpret, explain or analyse the ethnographic phenomena with which one is confronted. It seems like a version, in other words, of the standard worry of whether it is even possible to take off the socially, culturally, politically (etc.) 'tinted glasses' through which we must necessarily see the world, which typically in anthropology is designated technically as 'ethnocentrism' (see also Argyrou 2002).

However, what makes the ontological turn distinctive is the fact that it fundamentally recasts and radicalizes this problem by exploring the consequences of taking it to its logical conclusion. The epistemological problem of *how one sees things* is turned into the ontological question of *what there is* to be seen in the first place. Accordingly, what ultimately tints the anthropologist's glasses are not social, cultural, political or other presuppositions, but ontological ones, by which we mean basic commitments and assumptions about *what things are,* and *what they could be* (including things like society, culture, politics and power). Here, longstanding epistemological worries about ethnocentrism, solipsism, essentialism, orientalism and so forth are reconceived as ontological problems: How do I, as an anthropologist, neutralize or otherwise hold at abeyance or

5

in continuous suspension my assumptions about what the world is, and what could be in it, in order to allow for what is in my ethnography to present itself as what it is, and thus allow for the possibility that what is there may be different from what I may have imagined? The ontological turn is not so much a matter of 'seeing differently', in other words. It is above all a matter of seeing *different things*.

Hence the flagship term, 'ontological', indicates the need to shift anthropological concern onto questions about what kinds of things might exist, and how. But the notion of a 'turn' is also more than mere rhetoric in this context. Certainly, the term is meant partly to advertise as novel its response to basic questions of anthropological methodology, as is the case with other self-purported 'turns' in recent social theory – linguistic, ethical, affective and so on. More importantly, however, the notion of a turn in this case also describes the particular modus operandi that this methodological reorientation implies, drawing attention to the basic *reversal* involved in understanding the problem of tinted glasses as an ontological one. For if solving this problem has always involved finding ways to question or otherwise qualify presuppositions that stand in the way of 'grasping the native's point of view', to use Bronislaw Malinowski's original formulation of the anthropological challenge (1961: 21), thinking of these presuppositions as ontological implies a radicalization of this quest, such that anthropologists' capacity to 'turn' their own presuppositions – and thus to transform their field of analytical perception – is released to its maximal potential. The signature move of the ontological turn is just that: a thoroughgoing attempt to turn on its head the relationship, as well as the hierarchy, between ethnographic materials and analytical resources. Rather than treating ethnography as the object of analytical concepts and procedures, the turn to ontology treats ethnography above all as their source. To return to our opening example, Maori gifts and the spirits they are deemed to contain are treated, not as the effects of 'collective representations', as per Durkheim and Mauss's own sociological theory for example (1963), but rather as an analytical starting-point from which to rethink what, say, a 'collective', or a 'representation' for that matter, might *be* in the first place.

At stake, then, is a basic reversal from striving to grasp 'the native's point of view', to finding ways to overcome what one already grasps in order to better be grasped *by it* – and that's all 'the turn' is! As we shall see throughout this book, however, this basic move has profound consequences for how we think about the whole project of anthropology, including its basic modi operandi and methodological wherewithal, as well as its political ramifications and critical potentials. Questioning the authority of elementary contrasts that are often presented as foundational to the project of anthropological research (between, say, nature and culture, individual and society, matter and symbol, and indeed data, method and theory), the ontological turn elevates the contingencies of ethnographic materials as a platform from which to refigure the activity of anthropology itself, in a spirit of abiding empirical, theoretical and methodological experimentation. In this process, core objects of study (exchange, kinship, personhood, ritual, artefacts, politics), theoretical debate (e.g. society, culture, time, belief, materiality, power, subjectivity), and methodological concern (e.g. data, evidence, comparison, generalization, model making, research ethics) are all rendered open to wholesale reconceptualization. What are the objects and forms of anthropological inquiry, and what could they become through exposure to the contingencies of ethnography, are the irreducibly ontological questions that lend the 'turn' its name.

It is important to note here that the empirical material that occasions such reconceptualizations can be drawn from anywhere, anytime, and by anyone, for there is potentially no limit to what is amenable to ontological analysis and critique. A mistaken (if partly understandable) consensus has taken root within certain quarters of anthropology that only particular questions, themes and topics, as well as (even more problematically) particular peoples and places lend themselves to, or are even 'worthy' of, the kind of analysis and thinking the ontological turn provides. While in

7

Introduction

the chapters that follow we shall see that some of the most decisive steps in the development of this line of anthropological thinking emerged from studies conducted in such 'traditional' ethnographic locations as Melanesia and Amazonia, and in relation to such classic anthropological topics as ritual, gift exchange and animism, in principle, and increasingly in practice, there is no limit to what discourses, practices and artefacts are amenable to the approach of the ontological turn. What might seem an anachronistic if not downright dangerous theoretical approach applicable only to 'indigenous cosmologies' and 'tribal' or 'non-Western' peoples, can and should be extended to all sites, themes and questions, including, in some of our own recent work, such 'hardnosed' political problems as security, revolution and empire (e.g. Pedersen 2011; Holbraad & Pedersen 2012, 2013; Pedersen & Bunkenborg 2012; Holbraad 2013b). Other recent works that adopt an ontological approach, often elaborating upon it critically in innovative ways, include studies of such diverse topics as money (Maurer 2005, Holbraad 2005), healthcare (Kelly 2011), transnational migration (Elliot 2016), medical anthropology (Bonelli 2015), architecture (Corsín Jimenez 2013; 2014), postcolonial land reform (Nielsen 2011, 2014; Di Giminiani 2013), new social movements (Krøijer 2015; Heywood 2015); infrastructure (Jensen & Winthereik 2013), new public management (Ratner 2012); creativity (Hirsch & Strathern 2004; Leach 2014), fashion (Vangkilde 2015), contemporary music (Born 2005; 2010), climate change (Hastrup 2011), games and calculation (Pickles 2013), natural science and natural scientists (Candea & Alcayna-Stevens 2012; Helmreich 2012; Walford 2015) and digital worlds (Knox & Walford 2016; Boellstorff 2016; Hogsden & Salmond 2016).

Still, as our opening examples illustrated, one of the central messages of this book is that there is nothing inherently new in the ontological turn. Rather than a radical rupture from the anthropological past, we suggest, the turn to ontology with which we are concerned here is oriented towards releasing in their fullest form potentials that have always been at the heart of the discipline's intellectual project, and that are

exemplified in many of the greatest exponents of that particular form of thinking we call anthropological. While not pretending to provide an exhaustive intellectual history, in the chapters that follow we trace some of these trajectories of anthropological thought. As we shall see, reconstructing the intellectual genealogy of what eventually became anthropology's ontological turn involves examining certain developments within, and traffic between, its three so-called main traditions, represented in our earlier examples, namely the American, the British and the French, personified in the works, respectively, of Roy Wagner, Marilyn Strathern and Eduardo Viveiros de Castro, which form the core of the theoretical lineage we seek to articulate in this book.

So if the ontological turn is not meant as a revolutionary rupture with anthropology's past but rather as a continuation of some of its most distinctive traditions, then where does its originality lie? Over the following pages, we show how the most distinctive contribution of the ontological turn consists in the way in which it systematically deepens or 'intensifies' existing but partly dormant potentials in the anthropological project. More precisely, we contend, the turn to ontology involves deepening and intensifying three abiding modes of anthropological thought: reflexivity, conceptualization and experimentation. We call these the three 'ontological turnings'.

Three Ontological Turnings

Reflexivity: The ontological turn's radicalization of anthropologists' longstanding commitment to 'reflexivity' is an obvious place to begin. After all, the easiest way to grasp the significance of what we have called the basic 'reversal' marked by the ontological turn – that of giving logical priority to the ethnography over its theorization, in order to release its full potential as a source rather than just an object of anthropological thinking – is to think of it as a particular manner of intensifying the call to reflexivity in anthropology. In the broadest and most inclusive sense, one may think of the

9

call to reflexivity as the injunction, in whatever one is doing, to be attentive also to the manner in which one does it – its conditions of possibility, so to speak. The basic move of the ontological turn in this connection is as simple as it is profound: yes, focus reflexively on the conditions of possibility of anthropological knowledge; but think of these conditions ultimately not as social, cultural or political, but as ontological ones – which is to say, conditions pertaining to what things might be.

It is important to stress here that, in the context of this argument, 'the ontological' does not refer to some kind of substantive level or field of phenomena – one, say, that might be distinguished from other such levels or fields (e.g. social, cultural, political, moral, aesthetic, economic, mental, biological, affective) mainly in being somehow 'deeper' or more 'fundamental' than them. This being, presumably, the shadow of a vaguely philosophically derived notion of ontology as concerned with the deepest level of existence, pertaining to grave matters of Being, foundational categories and so on. As we shall see in the next chapter, some anthropologists who have been appealing to the notion of ontology in recent years have taken it in this 'deep' sense, while others have committed themselves to a full-scale metaphysical revision of the world's make-up inspired by recent developments in continental philosophy as well as Science and Technology Studies (STS). However, in the way we seek to expound it here, if anything is 'deeper' about the ontological turn when compared to standard forms of social, cultural, political or other reflexivity, that is the manner in which it enacts the call for reflexivity itself. And this is not because ontology is taken to mark out some more solid, and in that sense 'deeper', level of reality that might encompass or otherwise ground other fields (social, cultural, political, or what-have-you) imagined as more derivative or shallow than it. On the contrary, to pose the question of anthropological assumptions in ontological terms – to ask, what kinds of things are there? – is above all to *refuse* to take as axiomatic any prior commitment as to what kinds of things might provide the ground for a reflexive turn in the first place (e.g. society, culture, politics and so forth).

keep open question of what might be

So, operating always as an adjective or adverb – never as a noun! – 'the ontological' here is meant as a call to keep *open* the question of what phenomena might comprise a given ethnographic field and how anthropological concepts have to be modulated or transformed the better analytically to articulate them. To take the ontological turn is to ask *ontological* questions without taking *ontology* as an answer. It is in this sense that it represents an intensified, more thoroughgoing commitment to more traditional forms of anthropological reflexivity, rather than a rupture with them. Instead of closing off the horizon of reflexivity in the name of some sort of ultimate reality that may ground it (an 'ontology' in the substantive sense), the ontological turn is the methodological injunction to keep this horizon perpetually open. To recall Clifford Geertz's (1973: 28–9) invocation of the old adage about turtles (though he of course was writing about culture), this intensified manner of reflexivity *goes all the way down*.

This may sound like an entirely debilitating way of thinking. It conjures an image of anthropologists forever mired in an inward-looking self-critique, unable to say anything positive about the ethnographic worlds with which they engage – this being the standard charge against anthropology's so-called crisis of representation in the mid-1980s (e.g. *crisis of rep.* Marcus & Clifford 1986). Isn't all this just a new version of postmodernist anthropological 'navel-gazing', only worse for being so deliberate and, seemingly, conservative? In a manner that may at first appear paradoxical, however, avoiding just this pitfall is very much part of the point of the 'turn' to ontology. To adapt a metaphor from Roy Wagner (1981), the ontological turn involves a 'figure/ground reversal' of the very idea of reflexivity, such that ethnography becomes the ground against which ontological commitments – what is x? – are figured and refigured (see also Box 0.1). For if, in its postmodernist version, anthropological reflexivity took the form of 'deconstruction' – critically debunking positive representations with reference to the hegemonic (social, cultural, political etc.) conditions of their production – then, in its ontological version,

figure ground reversal

Introduction

reflexivity turns the critical energy of deconstruction into a positive agenda for generating – *constructing* – new ways of thinking.

The change really is quite simple, yet has far-reaching consequences for the anthropological project: if what gets in the way of seeing new things in our ethnography are prior ontological assumptions as to what those things can be in the first place, then overcoming this predicament of ontology (to paraphrase Clifford on 'the predicament of culture' – 1988) must involve making those assumptions explicit, and then changing them. So the very requirement (critically) to move away from one set of assumptions precipitates the need (positively) to refigure them in a way that allows previously obscure aspects of the ethnography to become apparent. It is in this sense that the ethnography becomes the ground of new concepts, providing the lever with which anthropological perception can be transformed. By radicalizing anthropology's call for reflexivity to the point of reversal, then, asking ontological questions in this way 'turns' the negative procedure of deconstruction into a positive procedure for re-construction.

Box 0.1: Why the ontological turn is not relativism

For a discipline used to plotting the differences between theoretical positions on an axis running from universalism to relativism, the ontological turn can easily feel like an extreme form of the latter. Certainly, if relativism involves a tendency to 'relativize' things that might otherwise seem absolute, then the ontological turn seems to intensify relativism in the same sense as, we argue, it intensifies anthropology's commitments to reflexivity, conceptualization and experimentation. Relativists tend to relativize by pointing out the ways in which forms of knowledge, truth or morality are contingent on differing social, cultural or historical circumstances. The ontological turn takes this to its logical extreme by questioning (note: this is not the same as denying) the universal validity of *everything*, including

---

such notions as knowledge, truth, morality, society, culture and history – the very concepts, that is, used to set up arguments also for (or against) relativism.

It is just this intensified commitment to anthropological relativization, however, that makes the ontological turn fundamentally different from relativism. The difference comes down to where the act of relativization is *located* in either case. Relativism imagines the world as inherently differentiated into different social groupings, cultural formations or historical moments, and then 'localizes' varying claims to knowledge, truth or morality with reference to them. The standard act of relativization, then, pertains to the relationship between the ethnographic data in which anthropologists are interested and the varying social, cultural or historical 'contexts' to which these data belong.

For the ontological turn, by contrast, relativity is located not in the relationship between ethnographic data and varying social, cultural or political contexts, but rather in the relationship between the variable data in question and the varying ontological assumptions anthropologists must inevitably make in their attempt to describe them ethnographically. This is because the very terms anthropologists use to describe their data have built into them particular assumptions about what these data *are*, and these ontological assumptions, which carry over into their subsequent analyses, are themselves contingent. So the intensified relativity of the ontological turn lies in making these basic ontological implications of all anthropological knowledge production prime objects of analytical attention. In particular, the act of relativization here pertains to the relationship between variable *objects* of description and varying *terms* of description. Relativizing this relationship is necessary because the terms anthropologists use to describe their data, and the ontological assumptions they inevitably make in doing so, may well turn out to be inappropriate, producing imprecise, inconsistent, incongruous or otherwise inadequate descriptions and analyses. The anthropologists' task, then, must be to *shift* the contingent ontological assumptions that render their initial ethnographic intuitions and descriptions inadequate, in order to arrive at concepts that will allow them to describe and analyse their ethnographic data more cogently and precisely.

To return to our introductory example: Say we describe as 'gifts' the items Maori call *taonga* (see Weiner 1992: 165 for comments on the translation). One of the things we might assume at the outset about gifts in general is that they are a kind of 'object' (e.g. an object that is exchanged). But then the indigenous account Mauss provides has these Maori gifts as, famously, containing the spirit of the donor. So in what sense can the *taonga* be described as 'gifts', if that involves assuming that they are objects (as opposed, for example, to subjects, or perhaps something in between)? Even just describing *taonga* as gifts is skewed, preventing us from articulating consistently *what these things are*. Therefore, the move of the ontological turn here is to relativize the anthropological concept of 'gift', reconceptualizing it in relation to, among other things, the analytical distinction between subjects and objects, so as to arrive at a new way of understanding what a gift might be. Crucially, such a new concept of the gift would have to avoid the initial confusion of assuming that something described as containing a spirit could be taken also as a kind of object.

Conceptualization: It is just this capacity, not only to subject one's prior assumptions to critical scrutiny, but to generate new kinds of, and instruments for, thinking out of one's ethnographic materials that defines the second sense in which the ontological turn intensifies the anthropological project, namely through the cardinal role it accords to the work of conceptualization. Indeed, conceptualization connects directly to the basic methodological and fundamentally critical thrust of the ontological turn, to the extent that 'concept' here should be read as more or less synonymous to the more grave-sounding expression 'ontological assumption'.[2] If an ontological assumption is an assumption about what something (including what 'a concept') is, then it depends also on how the concepts (including concepts of concepts) involved in articulating it are defined. To ask, for instance, What is a person? is to ask, How is a person to be defined?, which can be taken as the same as asking, How are persons to be conceptualized? So to assert, as the ontological turn does, that anthropologists' engagement with their ethnography may require a shift of their ontological assumptions, is to claim also that how to conceptualize things (including, again, concepts) in a given ethnographic encounter is among its most basic concerns.

Once again, then, more than a break with earlier ways of doing anthropology, this focus on conceptualization is better considered as a particular way of extending and intensifying aspects of anthropological practice that have been present in the discipline for a long time. Certainly, taken on its own, the idea that anthropological thinking may involve the need to revise one's concepts is hardly novel, considering that ideas about anthropology's role in questioning that which is taken for granted, relativizing things, denaturalizing them, displaying the variability of human ways of being, including ways of thinking, or of 'seeing the world', and so on, are so common that they almost appear banal when listed in this way. Indeed, anthropology's special gift for using peoples' varied lifeways to present alternatives to what we may otherwise have taken for granted is also what has lent the discipline its sharpest critical edge – its abidingly political mission of what is

---

[2] Again, this is not offered as a philosophical thesis about the relationship between ontological assumptions, concepts and definitions, much less as an attempt to delineate the proper remit of ontology or metaphysics. This is a task for philosophers (e.g. Honderich 1995: 634). Ours is only an attempt to articulate clearly how the particular manner of anthropological analysis in which we are interested in this book operates, this being a point of anthropological methodology *par excellence*.

Introduction

sometimes called 'cultural critique' (Marcus & Fisher 1986; Hart 2001), or what Foucault hailed as anthropology's role as a 'counter-science' (1970: 378). Seen in this light, our rather cerebral-sounding insistence on anthropology's capacity for concept creation may perhaps look meek and non-engaged in comparison. And yet, assimilating our call for conceptualization to such longstanding 'critical anthropological' concerns fails to recognize the ways in which these concerns are recast by the ontological turn. This is so for two interconnected reasons.

First, it should be noted that, more than just pointing out the (obvious) need for anthropologists to pay attention to their concepts, the ontological turn makes this the *pivotal* task for anthropological thinking – its primary challenge. Conceptualization, in this sense, is the trademark of the ontological turn just as, say, 'explanation' epitomizes positivist approaches and that of 'interpretation' typifies hermeneutic ones. Indeed, much of the theoretical traction of the ontological turn comes down to the alternative that it presents to this rather hackneyed choice in the social sciences, between explanation and interpretation. For anthropologists to imagine their task as that of explaining *why* people do what they do, they must first suppose that they understand *what* these people are doing. The ontological turn often involves showing that such 'why' questions (explanation) are founded on a misconception of 'what' (conceptualization). E.g. the question of why certain people might 'believe' in nations, say, or ghosts, may be raised precisely because questions as to what a nation or a ghost (and indeed what 'belief' and 'doubt') might *be* have not been properly explored. And similarly for hermeneutics: conceived as cultural translation, to imagine that one's job as an anthropologist is to 'interpret' people's discourse or actions one must assume that one is in principle equipped with concepts that may facilitate such a process. To this the ontological turn counterposes the possibility that the reason why the things people say or do might require interpretation at all may be that they go beyond what the anthropologist is able to understand from within his

16

conceptual repertoire. Once again, the task of conceptualization assumes paramount importance.

This brings us to our second point concerning the central role of conceptualization in the ontological turn. For while the contrast between conceptualization and explanation is stark, the distinction between conceptualization and interpretation may seem too fine to be clear. Does the call to conceptualization really add anything to the old insight that in order to understand what people say and do one might sometimes have to change the way one thinks oneself? Is classic anthropological interpretation not in fact *all about* articulating, and comparing, ethnographically alternative conceptual universes – what used to be called 'worldviews', 'local knowledge' or just 'cultures'? To be sure, there is a degree of continuity between interpretation and conceptualization. Once again, however, the difference lies in the particular manner in which the project of ontological reconceptualization that we are articulating here deepens and intensifies earlier ways of thinking about the role of conceptualization in anthropological analysis. For even if conceptual shifts have always been a feature of anthropologists' interpretations and comparisons, they have most typically exhausted themselves in insights of the type 'for the x (e.g. peasants), time is circular, with the past ever returning to the present', 'among the y (e.g. nationalists), the whole amounts to more than its parts', or, 'according to the z (e.g. animists), things have spirits' and so on.

As we shall explain in more detail in the chapters to come, it is in fact surprising that these kinds of propositions have been accepted for so long in anthropology as genuine possibilities for thinking – alternative 'conceptualizations' of time, totality, objects and so on. As with the very idea of cultural relativism that they are so often used to demonstrate, the fact that such statements are so familiar to anthropologists and their audiences as pronouncements about other people's 'beliefs' of 'worldviews' takes nothing away from the basic fact that in themselves they are ambiguous in the very least, and often deeply incongruous and confusing. To be sure, in anthropology we use such evocative shorthands all the

Introduction

time. Nevertheless the idea of, say, a past ever-returning to the present (e.g. Eliade 1991; cf. Gell 1992: 30–6) is in itself confused: what exactly is 'past' about the past if it can be said to be ever-returning to become present? Similarly, beyond its elegant air of paradox, in what sense exactly are parts that add up to less than the whole 'parts' at all (e.g. Bohm 1980; Durkheim 2006; Bubandt & Otto 2010)? And what is the idea of a 'thing' meant to amount to when we say that it 'has a spirit' (e.g. Mauss 1990; cf. Henare et al. 2007: 16–20)? Aren't 'things' precisely the kinds of things that do *not* have spirits, by definition?

The ontological turn distinguishes itself from such forms of relativism by taking seriously the work of conceptualization that they (should) imply. It starts from the premise that what makes genuinely 'alternative' the possibilities for thought that ethnographies can provide is that they can go beyond the anthropologist's capacity to describe them by (ab)using concepts in their familiar senses. To avoid the conceptual vagueness that such descriptions involve, and the confusions that they so easily create, ontologically minded analysis takes on the task of providing the conceptualizations that are needed to render ethnographic descriptions and anthropological comparisons fully articulate: how might one indeed conceptualize time as circular? What might past, present and future amount to in such a manner of thinking? What do we need to do to the notions of 'part' and 'whole' in order to alter the logical coordinates of their relationship? And what might a thing be, and what a spirit, for the two to be conjoined conceptually? In taking seriously the requirement to follow through with these kinds of conceptual experimentations, then, the ontological turn is ultimately an attempt to take the challenge of relativism to its ultimate conclusion.

Experimentation: The third way in which the ontological turn 'intensifies' the anthropological project lies in its commitment to experimentation. Once again, this should be understood as an intensification of ways of operating that already have deep roots in anthropology, rather than a

18

rapturous new beginning. In fact, there are at least two senses in which the practice of anthropological research has always been experimental. First, ever since Malinowski, anthropology has identified itself as an empirical science whose practitioners gather their data in 'the field' as opposed to 'the armchair' by way of the celebrated method of participant observation. And, while relatively little attention has been paid to this, anthropologists' abiding commitment to participant observation involves them using themselves as research instruments for registering their field-observations within a thoroughly self-experimental research design (see also Shaffer 1994). After all, when one thinks about it, to an even higher degree than for instance the discipline of medicine, whose 'history ... is replete with ancient and contemporary examples of doctors who chose themselves to be ... volunteers for research' (Kerridge 2003: 204), the discipline that we have come to know as modern anthropology relies on the fieldworker's ability, and willingness, to use his or her own body and mind as both an instrument and an object of investigation.

Second, it is widely agreed that scientific experimentation involves an element of 'manipula[tion] of phenomena in such a way that answers can be given to specific questions' (Honderich 1995: 262). To be sure, anthropologists would do much to avoid using the term 'manipulation' to describe what they do in the field. Indeed, they often like to remind themselves about the ethical and political commitments that grow out of long-term fieldwork, and perhaps even more so as the conceptualization of (and relationship to) the people they study has shifted from one of 'informants' who constitute the 'object of study' from which data are gathered, to 'interlocutors' or 'respondents' along with whom knowledge is co-created in collaborative and inter-subjective research processes. Nevertheless, to claim that anthropological research does *not* involve a more or less controlled *intervention* in the lives of the people studied, as well as the lives of the ethnographers who study them, would be to ignore one of the central lessons from the discipline's 'crisis of representation' debate. Namely, that anthropological knowledge is the result of particular fieldwork encounters and contingencies (Rabinow 1977), and that its status therefore cannot be evaluated with reference to the objectivist principles stemming from the positive sciences, but should rather be understood as irreducibly intersubjective (Clifford & Marcus 1986: 166; see also Hastrup 2004; Jackson 1998). In fact, is that not where the often hailed yet not fully understood 'magic' of participant observations lies? As a unique form of research experimentation, it combines ethically informed and sustainable experimental engagement with the people studied with self-experimental interventions in the person studying them, thus collapsing two experimental methods into a single fieldwork *habitus*. Anthropologists, after all, do not just use themselves as research tools for generating their data. They also use the inevitable transformation of themselves and their interlocutors that this encounter involves as a primary source of knowledge and insight in its own right.

It is this inherently self-experimental impulse of the anthropological project that the ontological turn pursues to its logical conclusion. This involves not just experimenting with what fieldwork and field may be, or with what might be a fieldworker and an interlocutor (that is, the kinds of questions that have been posed by the entire generation of 'post-crisis of representation' anthropologists). It also – and above all – experiments with what an anthropological concept and an anthropological theory might be. In heeding the lessons of postmodernist anthropology and taking its call for reflexivity and cultural critique to the extreme, the ontological turn allows for an experimental extension of anthropology's self-experimental habitus from the so-called ethnographic dimension of 'the field' into the so-called theoretical dimension of 'analysis' itself (which may be another way of saying that the ontological turn amounts to a sustained experiment with what a concept, and indeed an experiment, could be[3]).

---

[3] Note the deliberate play here on the double meaning of the verb 'to experiment' as both rigorous scientific method and open-ended exploration. Indeed, the notion of experimentation suggested here would seem to be fundamentally at

Of course, it may be objected that this aspiration is far from new, to the extent that anthropologists have always sought to experiment with their analytical language and the categories by which they describe the people they study. But, once again, this is just our point: there is nothing radically new about the ontological turn with respect to the three abiding modes of anthropological thought we have singled out. Rather, the contribution of the ontological turn lies in the sustained and systematic way in which it seeks to take on board these three analytical injunctions by pursuing them all the way to their limits; as well as the

odds with conventional understandings among scholars and laymen alike of what constitutes a proper experiment in the natural (and indeed the social) sciences. After all, is scientific experimentation not widely defined as the quest for repeatable results through the setting up of controlled experimental environments that allow for the reproduction of 'invariance' and 'limited variables' over time? And does our vision for an experimental anthropology not represent an exact opposite ideal by advocating for maximal *variance* and the production of singular, non-repeatable results? To a significant extent, yes; but there are nevertheless grounds for arguing that what we are proposing here may be said to be experimental also in a more scientific sense. For one thing, as Latour has pointed out, there is in fact 'little difference between observation and experiment. ... An observation is an experiment where the body of the scientist is used as instrument, complete with its writing device, the hand. ... It does not matter if [one] has an instrument ... or ... huge laboratory-like paraphernalia' (1990: 57). What is more, as Latour also points out, 'nothing proves that an experiment is a zero-sum game. On the contrary, every difficulty [encountered by experimental scientists] suggests that *an experiment is an event* ... and not a discovery, an uncovering, an imposition, a synthetic a priori judgement, the actualization of a potentiality, and so on' (1990: 65–6). In other words, although it may not appear to do so (not even to its practitioners), ethnographic fieldwork – including the criteria of evidence associated with it (Engelke 2008) – arguably is a distinct mode of (self-)experimental research practice. In fact, if we are to follow the historian and philosopher of science Hans Rheinberger, the anthropological method possesses several *advantages* when held up against other, more laboratory-based 'research systems' (to borrow Rheinberger's term), notably the capacity to perpetually remain, as he puts it, '*young*': (1994). After all, proposes Rheinberger, '[r]esearch systems ... are characterized by a kind of differential reproduction by which the generation of the unknown becomes the reproductive force of the whole machinery. As long as this works, the system so to speak remains "young." "Being young", then, is not here a result of being near zero on the time scale; it is a function if you will of the functioning of the system. The age of such a system is measured by its *capacity to produce differences that count as unprecedented events and keep the machinery going*' (1994: 68; emphasis added).

manner in which it seeks to combine them in various ways (by, say, treating reflexivity and conceptualization as primary objects of anthropological experimentation in their own right). By always experimenting with fieldwork, but at the same time also consistently reflecting on what the field, fieldworkers, interlocutors, ethnographic data, theories and, above all, anthropological concepts might be, the ontological turn can be described as a deliberate attempt to strike an optimal balance between purposefulness and purposelessness within anthropological knowledge production.

This is also why, it almost goes with saying, many ontologically informed anthropological analyses are likely to 'fail', if by failure we understand the inability of a given ethnographic account and anthropological analysis to abide fully to the largely tacit criteria of evidence that are so characteristic to the discipline (Engelke 2008). Indeed, how could it be otherwise, given its fundamentally heuristic (Henare et al. 2007; Holbraad 2012; Pedersen 2012a) nature? The success of the ontological turn should not only – and perhaps not even primarily – be measured against its capacity to produce ethnographic accounts that are in perfect tune with dominant anthropological 'persuasive fictions' (Strathern 1987b), although, when it does occur, such a seamless integration is only to be welcomed. Much as with other heuristic forms (Candea 2015; cf. Wimsatt 2007), the success of an ontologically informed anthropological experiment is a function of the degree to which it can remain faithful to – and conscious of – its own design, including the inevitable but nevertheless productive limitations of its heuristic form. As Matei Candea puts it, heuristics 'don't simply fail, they fail in regular and predictable ways' (2015). And much as with other scientific experiments that are not closed and hypothesis-driven but open-ended and exploratory, success here lies in the extent to which a particular heuristic framework and its more or less 'algorithmic' procedures have been pushed to their very limit, while the modifications inevitably resulting from this process of extreme testing are transparently accounted for and reflected upon as part of the operation,[4] which in turn may help to explain why the very features likely to make an ontologically informed analysis look like a 'failure' from the (less reflexively experimental) vantage of prevailing taste regimes, often turn out to be the same elements that indicate its accomplishment in more ontological terms. So, indeed, failure is an 'endpoint' (Miyazaki & Riles 2005). Not, however, because anything new and novel is automatically to be preferred over the well-tried and the conventional, but because a successful 'run' of the experimental onto-logical 'machine' is to be measured against the degree to which poten-tially useful concepts have been generated by this heuristic procedure, and more generally the extent to which this ontological experiment has explicated, problematized and improved existing ways of thinking. (Just to make this clear, even if it may seem unnecessary: this does not mean that such heuristically generated analytical methods are necessarily any less 'ethical', 'political' or indeed 'critical' than prevailing hermeneutically informed forms of anthropological thought, an issue to which we shall have the opportunity to return in the Conclusion to this book).

In sum, one might say that the ontological turn's intensification of anthropology's experimental condition stems directly from the way in which it transforms cultural critique into conceptual creativity without in so doing losing any of its critical edge and reflexive impetus. For what is that which we earlier defined as 'intensified' or indeed 'radical' reflex-ivity and conceptualization other than the logical extension of anthro-pology's self-experimental habitus from the so-called ethnographic

---

[4] According to Hans Rheinberger, successful scientific experiments 'produce results that by definition cannot be produced in a goal-directed way. Given such condi-tions, a research device has to fulfill two basic requirements. First, it has to be stable enough so that the knowledge which is implemented in its functioning does not simply deteriorate in the course of continuing cycles of realization. ... Second, it has to be sufficiently loosely woven so that in principle something unpredictable can happen' (1994: 70–1). This, we suggest, is precisely what the ontological turn seeks to do by extending the object of anthropological experimentation from the artificially circumscribed arena of 'the field' to include the subsequent anthropo-logical analysis.

realm of 'the field' into the so-called theoretical realm of 'analysis'? As we shall show in detail especially in relation to the work of Marilyn Strathern in Chapter 3, by treating what one might otherwise imagine as the empirical and theoretical 'stages' of anthropological research as formally analogous components of one single undifferentiated analytical procedure, the ontological turn takes the method of (self)experimentation to its necessary endpoint. Ontologically inclined anthropologists distinguish themselves by *rendering their own thoughts* (and therefore their own concepts) subject to the same degree – and ideally, the same kind – of experimental intervention as the people whose lives they study and engage with in their field sites, including their own life as ethnographic fieldworkers. Ideally, this is what allows the ethnographic contingencies that emerge from fieldwork to transform perpetually the very concepts that one uses to describe and analyse these ethnographic materials anthropologically – self-experimentation, that is to say, *all the way down too.*

An Overview of the Book

Having in this Introduction presented in a nutshell the key tenets of the line of anthropological thinking we identify as the 'ontological turn', the rest of this book is devoted to delineating its development within the discipline, showing how it relates and contrasts with other approaches, and exemplifying critically the kinds of insight to which it can lead. It is important to make clear from the outset that, while our exposition will be critical, at times exploring important ambiguities and inconsistencies both within and across the writings that we review, our overall intention is positive. Having, since our own years as graduate students in the late 1990s, thought about, taught and published on different aspects of the ontological turn, our aim, in line with the book's title, is to 'expose' what we take this way of thinking to be, in the sense of 'putting it on the table' as a viable anthropological approach that can make important contributions to the discipline at the present juncture. So, as well as descriptive, our task is programmatic. Amidst the plethora of voices advocating, debating, criticizing, and sometimes confusing the different senses in which considerations of ontology may or may not be relevant to anthropological thinking, in this book we seek to identify, clarify and critically develop *one* such line of thinking, in order to make the case for it: showing how it came about, how it works, what it can and can't do for anthropology, and why it is worth pursuing.

We begin, then, with a first chapter surveying the growing literature on the role of ontology in anthropology, connecting it also to debates taking place in related fields, including recent developments in philosophy as well as in STS. Here we identify four broad ways in which the turn to ontology has been discussed, all of which can be viewed as alternatives, in different senses and to different degrees, to the particular line of thinking to which this book is devoted. Two of them, as we shall see, correspond to particular trends within contemporary continental philosophy and STS, respectively, while the other two constitute tendencies within anthropology itself. Branding these 'other ontological turns', we clarify the ways in which they relate to and inform the approach we delineate in this book, but also the ways in which they contrast with it. The distinctively methodological (and for the same reason fundamentally anti-essentialist and non-metaphysical) injunction towards reflexive conceptual experimentation that characterizes 'our' ontological turn, as we outlined earlier, will be at the centre of this discussion.

Chapters 2, 3 and 4 together present the intellectual development of this orientation in a loosely chronological order, by examining respectively the ground-breaking work first put forward by Roy Wagner in the 1970s, followed by an account of Marilyn Strathern's important contributions from the 1980s onwards, and completing the exposition with a presentation of Eduardo Viveiros de Castro's work, which has taken the approach forward in the past couple of decades. As we shall see, the notion of ontology emerged as an explicit concern only late in this trajectory of thinking, particularly in the work of Viveiros de Castro, so that our account of its development involves exposing its roots in Wagner's and Strathern's manners of radicalizing anthropological reflexivity, conceptualization and experimentation along the lines we have already indicated.

Accordingly, in Chapter 2 we discuss in detail Wagner's theory of invention, including his model of 'obviation sequences' which was originally developed for the study of myth. While Wagner has little to say about ontology, we show that his sophisticated and systematic manner of putting questions of conceptualization at the heart of anthropology effectively laid down the theoretical tracks for what later developed into the ontological turn. In Chapter 3, we explore Strathern's radically reflexive method for experimental ethnographic description, epitomized in her rethinking of the role of relations, comparisons and scales in anthropological analysis. Emphasizing the importance of feminism in Strathern's anthropological project, we show that her characteristically cautious or 'hesitant' attitude towards 'big' questions (including, therefore, also ontological questions posed in their conventional metaphysical key) illustrates the intensely reflexive stance that lies at the heart of the 'ontological turn'. Finally, in Chapter 4 we show how Viveiros de Castro's explicit turn to ontology originated in a comparative, ethnological argument about 'perspectivism' and 'multinaturalism' in Amerindian indigenous cosmologies. Exploring in some detail his methodological concept of 'controlled equivocation', we also examine critically Viveiros de Castro's political argument for a 'decolonizing' anthropology that takes indigenous interlocutors 'seriously' – seriously enough, that is, to allow their manners of living to transform our manners of doing anthropology.

Our genealogy of the ontological turn in the work of Wagner, Strathern and Viveiros de Castro bears out our earlier point that this way of thinking has deep roots in the history of anthropology. In particular, we show how the approaches of these three authors take forward in a unique way central questions in the American, British and French traditions of anthropology, respectively. Wagner's notions of invention and obviation constitute a thoroughgoing transformation of the central concept of American cultural anthropology, to wit, that of culture. Strathern, on her part, can be seen as a direct heir of the British social anthropologists' (and in particular Radcliffe-Brown's) classic interest in social relations, which she turns into a profoundly reflexive and experimental project in her investigation of the possibilities and limits of this theoretical imaginary. Finally, Viveiros de Castro is a poststructuralist in the fullest sense, inasmuch as his theorization of perspectivism and multinaturalism deliberately actualize conceptual potentials that have always been present within the French ethnological tradition, and particularly the concern with transformation that lies at the heart of Lévi-Strauss's structuralism. Seen in this light, we contend, the ontological turn represents a peculiarly creative imbrication of the three grand traditions of anthropology, and by that virtue takes each of them forward in new directions.

The final chapters, Chapter 5 and Chapter 6, mark a change in the book's format, from exposition proper to a more explorative key, by examining the ontological turn's purchase on two themes of anthropological enquiry that have been the subject of increasing attention in recent decades, namely material artefacts and Christian faith. Chapter 5 connects the ontological turn with recent debates about materiality and 'posthumanism' in anthropology and other fields, in order to explore how far material objects might provide a platform for the same kinds of conceptual experimentation the ontological turn draws from its ethnographic encounter with people. Built on a critical review of attempts by a new generation of scholars (including ourselves) to transpose the central tenets of the ontological turn onto the study of 'material culture', the argument here centres on how far artefacts can dictate the terms of anthropologists' analytical engagements with them by virtue of their peculiar character qua 'things'. Based on a re-interpretation of our own work on Afro-Cuban divination and Mongolian shamanism, we ask how, in particular, the qualities commonly imagined as 'material' might make a difference to the economy of anthropological analysis (including to the very distinction between the material and the immaterial, for example). What, indeed, happens to the analysis of things if we treat them, in a pertinent sense, as concepts?

Chapter 6 explores the ethnographic and analytical limits of relationality as an abiding theoretical imaginary of contemporary anthropology generally and of the ontological turn in particular. Using as our point of departure recent criticisms of the so-called New Melanesian Ethnography (NME) associated with Wagner, Strathern and their followers, we explore what might come 'after the relation' by exposing this core concept of the ontological turn to the ethnographic contingencies of Christian conversion and worship, which, as the NME critics and others have suggested, seem to resist relational analysis in fundamental ways. Yet, we argue, for such a 'post-relational' move to be faithful to the ontological turn's impetus towards reflexive conceptual experimentation, it should not involve an attempt to roll back the relational analytics associated with Wagner, Viveiros de Castro and, above all, Strathern. On the contrary, asking what comes after the relation involves transforming the very concept of 'the relation' by exposing it to ethnographic phenomena that have not hitherto been part of its conceptual purchase, including, as in our two illustrations, the notion of divine transcendence associated with some forms of Christian faith and the converted Christian subject. Here, then, post-relational analysis involves using the concept of the relation as a point of departure for further transformation through reflexive analytical experimentation when confronted with ethnographic materials that seem to cut against it, such as the apparently non-relational features of certain aspects of Christian faith and practice.

Finally, in the Conclusion we take stock of the overall argument of the book, putting the ontological turn in broader perspective on the current landscape of anthropological theory. In doing so, we bring to the surface a question that runs throughout the chapters of the book, concerning the political implications of this way of thinking about anthropology, suggesting a distinction between three ways in which ontology and politics are correlated: the traditional philosophical concept of ontology, in which 'politics' takes the implicit form of an injunction to discover and disseminate a single absolute truth about 'how things are'; the sociological critique of such essentialisms, which, in debunking all ontological projects to reveal their insidiously political nature, ends up affirming sceptical debunking as its own ideal of 'how things should be'; and the anthropological concept of ontological reflexivity, in which possible forms of existence are multiplied in accordance with ethnographic variability, so that politics becomes the non-sceptical elicitation of a manifold of potentials for 'how things could be'. Viewed in this way, we suggest, the ontological turn re-invigorates a unique but slumbering tradition of anthropological critique that intervenes in the world by making visible the myriad ways in which persons and things can alter from themselves.

ONE

Other Ontological Turns

Ontology, some would claim, is the current zeitgeist in anthropology, as well as in a number of related fields in the social and human sciences. Certainly, the expression 'ontological turn' has been deployed in a whole array of contexts in recent years, gaining a much broader currency than the one the present book is about. So, having conveyed in the Introduction the signature features of the ontological turn that we hope to articulate in our exposition, the next step is to place this particular way of thinking in the broader landscape of contemporary anthropology and social theory at large. Accordingly, the task of this chapter is to review recent debates about ontology within anthropology and related fields. Our aim will be to give a sense of the numerous ways in which the particular line of anthropological thinking for which, in the chapters that follow, we shall very programmatically be reserving the tag 'ontological turn' is allied with other ways in which ontology has featured in these debates, and to clarify the specific senses in which we take it to be nevertheless distinct from them. Since it would be impossible here to chart the whole landscape of recent social theory, for purposes of our exposition we limit ourselves to discussing the fields in which the notion of an ontological turn has gained traction most explicitly and which are most contiguous to our concerns. In what follows, we focus on recent developments in philosophy, Science and Technology Studies (STS) and, especially, anthropology itself, within which we shall distinguish two broad tendencies that can be contrasted to 'our' ontological turn.[1] Making up a total of four, we playfully call these the 'other ontological turns'.

As we shall be showing, the basic continuity between these developments and the ontological turn, as we understand it here, lies in a common desire to present alternatives to the fundamental ontological premises of what Bruno Latour – one of the key figures – has called 'the modern constitution' (1993). Transfiguring some of the orientations that used to get debated under the banner of postmodernism, and often motivated by the perception of a total (political, economic, ecological) crisis of empire, capitalism and modernity at large, these theoretical developments are symptomatic of an even broader impulse to invent new ways of thinking about, intervening in and experimenting with the world. So new ontologies now abound, in the form of an array of novel conceptual vocabularies and aesthetics of thought: assemblages, affects, networks, multiplicities; posthumanism, multi-species and the Anthropocene, new materialism, speculative realism; emergence, vibrancy, intra-action; the para-, the off-, the pata-. Inasmuch as generating such new forms of conceptualization – new and often deliberately experimental baselines for thinking – is at the heart of these 'ontological turns', their sympathy with 'our' version is basic.

Still, as we are going to argue, the version of the ontological turn with which this book is concerned is also significantly different from the others. The difference comes down to the core point about the fundamentally reflexive character of 'our' version of the ontological turn, which, as we showed in the Introduction, consists in a thoroughly methodological

[1] Although the authors we shall be reviewing are mainly social and cultural anthropologists, we should note that a turn towards ontology has also been debated hotly within archaeology (Webmoor & Witmore 2008; Alberti & Marshall 2009; Holbraad 2009; Alberti et al. 2011; Harris & Robb 2012). Benjamin Alberti in particular has been making a sophisticated case for the role of ontological reflexivity in archaeological analysis, extending to this field the purchase of the kinds of arguments we seek to explore in this book (e.g. Alberti 2014a).

proposal for 'intensifying' certain trains of thought that lie at the heart of the anthropological project. By contrast, notwithstanding the virtue they often make of being experimental, provisional and partial, the four 'other' alternatives that we examine in what follows tend in one way or other to assume, albeit often only implicitly, that a turn to ontology must involve participating in, or contributing to, the traditional philosophical project of building a metaphysical account of the basic constituents of existence. In some cases, as we shall see, this aim is explicit and deliberate, with anthropologists and STS scholars joining philosophers in providing their own answers to traditional ontological questions about how best to conceive of the world, its constituents and the relations between them (here, various critiques of Cartesian dualism have been prominent). Accordingly, much of this writing takes the form of ever-novel metaphysical stories about what the world is and how it works – relations replacing entities, processes swallowing up essences, assemblages and networks co-opting subjects and objects, flows and stoppages usurping the metaphysics of presence, only to be trumped by objects that retire into themselves, and so on, in speculative recalibrations of the conceptual armoury of metaphysical thought.

In other instances, which come closer to our approach, similar moves of conceptual revision are made, not in the name of arriving at a better ontological image of what the world is 'really like', but rather as a matter of methodological expedience. As we shall show, however, these methodological arguments are often themselves grounded on prior ontological commitments of their own. For example, the characteristic methodological notion that the world with which, say, anthropologists or STS scholars are empirically concerned is informed by an underlying set of ontological principles, which vary from one set of local practices to the next, seems in much STS writing to be itself premised on an (at least implicitly) metaphysical claim about the inherent multiplicity of the world – some have called this thesis 'ontological pluralism' (e.g. Law 2004).

So, in the rest of this chapter we sift through the different positions that have been taken within recent debates about ontology in order to set out the broader coordinates for the approach we shall delineate in the following chapters, teasing out critically the elements that make it most distinctive. And what better place to begin than with the old mainstay of ontology, namely philosophy.

Philosophy and 'Object-Oriented Ontology'

The first of the four 'other' ontological turns we shall discuss is associated with the loose grouping of philosophers who are referred to as 'speculative realists' and the part-overlapping constellation of scholars doing 'object-oriented ontology', or 'OOO' as one of its proponents, Ian Bogost, has branded it for short (2012: 6). Here, traditional philosophical and ontological questions such as, 'what kinds of things exist?', 'what makes something a world?' or, indeed, 'what is Being?' are probed to arrive at an authoritative metaphysics true to the world. As we shall see, such a characteristically philosophical orientation is in stark contrast with the decidedly reflexive and anthropological approach we seek to delineate in this chapter and in the ones to come. Nevertheless, what the speculative realists and OOO do share with other recent turns towards ontology in anthropology and related fields is scepticism towards the increasing sidelining of ontological questions that has arguably taken place within philosophy and the human sciences over the last decades, if not centuries. What unites the diverse thinkers associated with speculative realism and OOO is thus a shared dissatisfaction with what they lament as the strictly epistemological (anti-metaphysical) path pursued by philosophy since Kant's *Critiques*, and a willingness to pose ontological questions about what exists anew, without thereby reverting to the antinomies, impasses and other dead ends of traditional dogmatic metaphysics. As Quentin Meillassoux writes in *After Finitude* (one of the most influential books in the field):

[C]ontemporary philosophers have lost the *great outdoors*, the *absolute* outside of pre-critical thinkers . . .; that outside which was not relative to us, and which was given as indifferent to its own givenness to be what it is, existing in itself regardless of whether we are thinking of it or not; that outside which thought could explore with the legitimate feeling of being on foreign territory – of being entirely elsewhere . . . (2008: 7; emphases original)

There are several reasons for which this return to the concerns of pre-Kantian philosophy is considered necessary for the speculative realists. Chief among these, asserts Graham Harman, whose own Whitehead-inspired theory of objects has spurred a lot of the recent surge in OOO literature (e.g. Morton 2013), is the fact that 'relationality [has become] a major philosophical problem. It no longer seems evident how one thing is able to interact with another, since each thing in the universe seems to withdraw into a private bubble, with no possible link between one and the next' (Harman 2010: 157). Once again, we recognize here the doggedly metaphysical notion that there is a 'really real' reality 'out there' – one that includes objects imbued with a transcendent depth and interiority, which is forever hidden to humans due to the fact that we can only interact with these objects by relating to them. Notwithstanding his differences with Harman, this is also the central question that Meillassoux asks, namely how to break free from what he refers to as the 'correlationalist circle' (2008: 53). The question for Meillassoux is to escape the Copernican shackles in which, as he writes, 'to be is to be a correlate'. The speculative realist problem, then, 'consists in trying to understand how thought is able to access the uncorrelate . . . whose separateness from thought is such that it presents itself to us as a non-relative to us, and hence as capable of existing whether we exist or not' (2008: 29). In their attempts to address these and similar metaphysical questions, the speculative realists and their OOO brethren cast their nets in disparate theoretical directions, ranging from Meillassoux's own Badiou-inspired philosophy of nature as radical contingency to

Harman's mix of Heidegger's tool-analysis with Bruno Latour's concept of the network (Latour 2009).

The key point to note for present purposes, however, is that these diverse approaches and perspectives share a straightforwardly philosophical ambition to provide the outline of a new metaphysics and, indeed, an alternative ontology – a single conceptual framework that would account, once and for all, for everything that exists. Granted, there are aspects of these metaphysical proposals that resonate with ideas that have been put forward by scholars associated with the ontological turn in anthropology and STS – e.g. Harman's peculiar but sophisticated theory of objects is explicitly allied to some of Latour's work (Harman 2009). Still, these philosophical goals differ fundamentally from the ones with which the present book is concerned. To be sure, we share with the OOO debate and the speculative realists more generally the desire to ask the 'forbidden' ontological questions that modern philosophy has for centuries taught us not to pose (and anthropology, sociology and other social sciences have tended to toe the line in this). And certainly, we remain sympathetic to the wider theoretical agenda of these recent philosophical trends, including the attempt to undermine and do away with the humanist concept of the *anthropos* and its 'intersubjective solipsism' (Meillassoux 2008: 51) via experimental explorations into 'what it is like to be a thing' and other 'alien phenomenologies' (Bogost 2012; see also Chapter 5).[2] Nevertheless, as Casper Bruun Jensen puts it in his own critical commentary, while on the face of it Meillassoux's 'argument is [an] apparent replication and intensification of anthropological

[2] Indeed, attempts have been made by anthropologists to draw on insights from speculative realists to grapple with particular ethnographic phenomena and derived ontological problems, even if these have invariably involved a fundamental departure from the more dogmatic and therefore, by definition, un-anthropological metaphysical assumptions mentioned earlier. See, for example, Pedersen (2013a), Jensen (2013) and Viveiros de Castro (2015). While Viveiros de Castro only alludes to the speculative realists in passing and Jensen is unequivocally critical, Pedersen's

ontologists' attack on culturalism, … Meillassoux's project runs directly counter to the ontological turn in anthropology' (2013: 327; see also Graeber 2015: 23).

In sum, whereas the speculative realists, in their role as metaphysicians, seek to formulate philosophically bulletproof conceptual frameworks that articulate true ontologies, what we care for as anthropologists is subject to a very different (since inherently contingent and always changing) control, namely that defined by the specificities of our ethnographic materials. Indeed, broadly understood, this point could be scaled up to the relationship between the ontological turn in anthropology and philosophical approaches more generally. It is true that, as it is often remarked by commentators as well as by some of its proponents themselves (e.g. Holbraad & Pedersen 2009; Viveiros de Castro 2009; 2014; Laidlaw 2012; Salmond 2014), the emphasis that the ontological turn places on questions of conceptualization, and the virtue it makes of keeping ontological horizons open and contingent, owes much to the influence of Gilles Deleuze and Felix Guattari, for whom philosophy is famously to be defined as the creation of concepts (Deleuze & Guattari 1994; see also Jensen & Rödje 2009). In Chapter 4 we shall see in relation, particularly, to the work of Viveiros de Castro some more specific ways in which the ontological turn can be articulated in the light of (in fact, in alliance with) Deleuzian conceptions. Nevertheless, one of our

'post-relational' analysis represents a more substantial and, up to a point, positive engagement. More precisely, his discussion of speculative realist ideas in an analysis of Mongolian ethnography serves a doubly critical purpose. On the one hand, he makes recourse to concepts from Meillassoux and Harman to describe aspects of his ethnography that cannot be captured by conventional relational analytics of the sort that is commonly associated with the ontological turn (as we are going to show in detail in chapters to come), namely the existence of occult phenomena held to exist 'outside' the otherwise all-encompassing shamanic cosmos. But, on the other hand, far from accepting the speculative realist metaphysical position, Pedersen uses these specific ethnographic contingencies to extend (and therefore criticize) not just conventional relational analysis but also the non-relational assumptions that underwrite Meillassoux, Harman and their peers.

central claims in this book is that, notwithstanding its affinities with Deleuzian philosophy, the ontological turn's own concern with the confection of concepts is born of exigencies that are peculiar to anthropological research and above all, as already intimated in the Introduction, the particular demands ethnography places on anthropologists' attempts to describe and analyse it.

To delve further into this distinction between metaphysical model building and the ontological reflexivity that is characteristic of the ontological turn's methodological orientation – asking 'ontological' questions without taking 'ontology' as an answer, as we put it in the Introduction – we now turn to the second of our 'other' ontological turns, which has taken place in the field of STS.

Science and Technology Studies

In their introduction to a collection of articles devoted to the rising prominence of ontology-talk in STS in the past couple of decades, Steve Woolgar and Javier Lezaun (2013) connect this development to the field's constitutive desire to present an alternative to the study of 'epistemology' in the philosophy of science. The ontological turn in STS, they argue, is best understood as the discipline's latest 'deflationary tactic' directed, not at providing 'more satisfactory answers to old epistemological questions, but rather to displac[ing] the framework that accorded them their central, obtrusive quality' (2013: 322). In particular, it is motivated by 'a desire to avoid being caught up in the description and qualification of "perspectives". It is an effort to circumvent epistemology and its attendant language of representation in favour of an approach that addresses itself more directly to the composition of the world' (2013: 321–2).

The recent turn in STS, then, has been 'ontological' in two related ways. First, it stems from a desire to sidestep a particular ontological framework, seen as particularly domineering inasmuch as it provides the foundations for the very idea of epistemology. According to this framework, the object of science is a single, uniform world 'out there' ('nature'), that constitutes the object of the 'perspectives' that human beings may take upon it ('cultures', 'representations', 'theories', 'discourses', 'epistemes' etc.), which are multiple since they vary from person to person or group to group, change at different times, and so on. Famously, drawing on Shapin and Schaffer's (1985) story of how intellectual, technological and political shifts came together in the seventeenth century to give birth to it, Bruno Latour brands this ontological regime of one-nature-many-representations as 'the modern constitution' (Latour 1993). As we shall see, in the guise of 'one-nature-many-cultures', this ontological framework is also deemed as a prime obstacle to be overcome by certain versions of the ontological turn in anthropology, including the one we are seeking to articulate in this book.

However, STS's turn is ontological also in a second sense, which is perhaps more particular to the field (although as we shall see some proponents of the ontological turn in anthropology share it). For STS, if the modern constitution of nature versus representations is to be overcome, that is because it belies the way science and technology actually operate. In particular, it obscures the ways in which scientific and technological practices are party to the very constitution of the objects with which they engage (Pickering 1995; 2016). As enactments rather than just representations of the world, in other words, these practices are best conceived as sites in which particular ontological configurations and effects emerge, interact with each other and are constituted and transformed through different socio-material practices and infrastructures. Much of STS theory has been about analytically articulating such a possibility and devising ways to track it empirically. At the analytical level, a prime concern has been to distinguish the idea that science and technology enact the world from the familiar late-twentieth-century notion that the

world, including science and technology, is a series of 'social constructions' (see also Hacking 1999; 2002). To speak of social constructions
(or cultural constructions – for present purposes it is the same), STS
scholars argue, is simply to ratify the modern constitution by assuming that the variation of the world must be predicated on its social or
cultural representations; construction, in other words, as a matter of
perspective. On the contrary, for STS, if the world is constructed by
science, or any other human practice for that matter, that is because
human practices participate in its very constitution by transgressing the
putative ontological divide between them and the world, through which
the modern constitution would seek to 'purify' them (Latour 1993: 10–
11). Hence, in place of social constructionism, STS posit the constitution
of the world as an open-ended operation, performed variably through
the different, and highly complex and often conflictual, engagements
and interactions between all of the world's constituents. Human beings
may be important as part of these chains of interconnections and reac-
tions (Michel Callon and, most famously, Bruno Latour have called
them 'networks' – Callon & Latour 1981; Latour 1993; 1999; 2005; cf.
2013: 30–46), but their operations and 'agency' can be understood only
in relation to a heterogeneous plethora of non-human entities ('actors'
or 'actants' – Latour 2005: 54–5): spirits, animals, ideas, institutions,
politics, laws, technologies, techniques, materials, artefacts, organisms,
molecules – whatever works, i.e. whatever makes a difference to the
particular practice in question. 'Hybrid' all the way down, these networks are 'symmetrical' in that they refuse to obey any principled divide
between humans and non-humans, just as they transgress the corollary ontological divides between representation and world, culture and
nature, and so on (Latour 1993; cf. 2005).

The upshot is an image of a world always in the making, emergent,
incomplete and as fragile and mutable as the practices and processes
that bring it into being and hold it, precariously, together: a world 'in
composition' (Latour 2010b; Woolgar & Lezaun 2013: 322). Indeed, the

more radical consequence of this is also often put forward: namely, that, contrary to the 'mononatural' ontology of the modern constitution, a world thus composed, which is to say a world that is always recomposable, is best conceived as *multiple* (e.g. Mol 2002). Differently configured localized practices will have different ontological effects, bringing forth different kinds of entities, with no overall ontological scheme to sort them out in a unified way (Jensen 2004; Gad & Jensen 2010). In such circumstances, it makes more sense to speak of the world in the plural. As enacted in tourist visits, for example, the famous landmark in the Australian desert, Ayers Rock, is something altogether different than Uluru, as Aborigines call the rock when they make it the focus of particular ritual practices and mythical narratives: the 'it' in question is two (in fact more) different things (Law 2004: 122–39). Indeed, as Law shows with reference to this politically fraught example, such a 'multiple worlds' or 'multiple ontologies' thesis has profound consequences for how we think also about politics, since it implies that, more than just a clash of opinions, views, beliefs or convictions, political dispute is a tussle about the very constitution of the world – a matter of 'political ontologies' (e.g. Verran 2014), 'ontological politics' (e.g. Mol 1999) or even 'cosmopolitics' (Latour 2004a; Stengers 2010). We shall see that, often adopting this terminology, in anthropology too politics features prominently in debates about ontology.

Here, however, we may note that STS scholars have developed a rich conceptual vocabulary in order heuristically to track and empirically to unpack the diversity of these ontological operations. Indeed, the sometimes perplexing quality of the neologisms of STS corresponds to its departure from the modern regime of ontological 'common sense', effectively inventing a new language to speak about possible alternatives to it. So alongside the by now standard Latourian vocabulary of networks, hybrids and actants, formalized as 'Actor Network Theory' ('ANT'), we have a growing mass of STS-inspired terminology: cyborgs and naturecultures (Haraway 1991), method assemblages (Law 2004), inscription devices (Latour & Woolgar 1986), transitivities (Law 2000), irreductions (Latour 1988), mangles (Pickering 1995), practices (Mol 2002), entanglements (Barad 2007), pleats (Bowker 2009), infrastructures (Jensen & Winthereik 2013; Jensen in press) and more. Even more than with OOO, this exuberant proliferation of analytical concepts and procedures is in profound sympathy with the project of reconceptualization that is so central to the ontological turn in anthropology as we understand it, and the flow of inspiration between these two currents of thinking has been mutually rewarding and often productive (see, for example, Strathern 1996; Jensen & Rödje 2009; Latour 2010a; Jensen et al. 2011; Gad, Jensen & Winthereik 2015; Jensen 2012; Jensen & Morita 2012; Lien & Law 2011, 2012; Blaser 2014; Jensen 2014; Viveiros de Castro 2014; de la Cadena et al. 2015; Walford 2015; Pickering 2016). But the important contrast with the speculative realists and OOO is that in STS these acts of conceptualization are not typically intended as the building blocks for new ontological frameworks with which to replace the ontological assumptions of the modern constitution. Rather, they are intended as conceptual devices with which to rescind these prevailing assumptions in order to allow the contingent possibilities engendered by worlds in the making to emerge. Woolgar and Lezaun put it sharply:

Having developed its characteristic analytical sensibilities in a series of moves of deflation and deflection, it would be odd if STS were now to embark on a project to champion one or another version of ontology. Instead ... the turn to ontology in STS can be better understood as another attempt to apply its longstanding core slogan – 'it could be otherwise' – this time to the realm of the ontological. (2013: 322)

In terms of our foregoing discussion, then, what distinguishes the ontological turn in STS from its manifestation in recent philosophical trends is its thoroughly reflexive and methodological quality. As with the form and method of anthropological thought to which this book is devoted, it is intended not as a theory but as a manner of proceeding – indeed a manner of experimentally disrupting the ontological status quo (cf. Lynch 2013).

Among the range of versions that this project has taken in the STS literature, some have been particularly deliberate on this point, presenting their analyses as reflexive attempts to experiment with novel ways of conceiving varying empirical materials (e.g. Winthereik & Verran 2012; Morita 2013; Jensen in press), much as we wish to do for anthropology in this book. At the other extreme, admittedly, there are influential STS scholars who come much closer to the spirit of the kinds of philosophical arguments we reviewed earlier, presenting their analyses as positive ontological proposals in their own right, which are truer to the world than existing – typically 'modern', 'Cartesian' – ones (e.g. Barad 2007; Pickering 2016; see also DeLanda 2002; 2006).[3] Perhaps the bulk of the STS literature, however, occupies a somewhat ambiguous position in between these two poles, often mixing an emphasis on methodological openness with a tendency, nevertheless, to foreclose its own experimental orientation in favour of a fairly uniform-looking image of what the world is like, and how it works – a world of practices and enactments, comprising assemblages, networks and recursive circuits, acting upon each other in myriad processes of mediation and translation.

Notwithstanding the basic confluences already mentioned, this tendency may indicate that there is also a significant divergence between the ontological turn, as we understand it here and the STS approach in at least some of its versions. The issue comes down to the role reflexivity

---

[3] To take just one indicative example of the approach we have in mind, consider how, in the context of a review of the ontological turn in STS, Andrew Pickering presents his 'own' alternative to what he calls representational idioms of scientific practice: 'The world – humans, nonhumans and whatever – just is an indefinite multiplicity of performative entities endlessly becoming in decentred and emergent dances of agency. This is the ontological picture I want to dwell on.... My ontology is a symmetric one of a multiplicity of reciprocally coupled emergent agents, human and nonhuman' (2016: 4-5, 6, footnotes omitted).

plays in either case and, particularly, how it is motivated. As we saw in the Introduction, our version of the ontological turn gives cardinal importance to reflexivity because it sees it as a prime condition of possibility for the very act of anthropological description. Reflexivity, on this view, is not grounded in some prior claim about the nature of the realities anthropologists countenance, but is rather seen as a necessary characteristic of the act of anthropological description itself. Simply put: in order to describe things with which we are not already familiar, we must attend reflexively to the concepts we use, and be prepared experimentally to reconstitute them. By contrast, it would appear that for many STS scholars the need for conceptual reflexivity stems in the first instance, not from the methodological exigencies of description, but rather from a prior set of understandings about the nature of the world (or worlds), and the demands it places on their own attempts to get a handle on it. In particular, the characteristic methodological openness and conceptual reflexivity of much STS scholarship is often grounded on a specific image of the ontological constitution of the world, according to which the world is *itself* open and in a state of permanent reconstitution. Thus, if STS scholars have found it necessary to experiment with new conceptual vocabularies, often this is because they deem the objects of their descriptions – the world or worlds – to be of a kind that requires such reflexive analytical moves. In such a case, however, what is presented as an essentially methodological, reflexive and constitutively disruptive ontological argument, ends up being pressed into the service of an authoritative metaphysical story about what the world is like, and how it works. A methodological argument is performed in the service of an ontological one, rather than, as 'our' ontological turn would have it, the other way around.

Making a new departure for his own intellectual project in a book tellingly titled *An Inquiry into Modes of Existence: An Anthropology of the Moderns* (2013), Latour himself comes close to admitting the problem.

Writing of an imaginary anthropologist trying to chart her subject matter (in this case 'the Moderns') using the tools of ANT, he remarks:

[A]s she studies segments from Law, Science, The Economy, or Religion she begins to feel that she is saying almost the *same thing* about all of them:, namely, that they are 'composed in a heterogeneous fashion of unexpected elements revealed by the investigation.' ... [S]omewhat to her surprise, this stops being surprising, in a way, as each element becomes surprising *in the same way*. (2013: 34; emphases original)

What is most telling, perhaps, is the remedy Latour offers for this predicament, namely an inquiry effectively devoted to charting out nothing less than the whole ontological constitution of modern society, modulating the ontological principles according to its 'segments' (Law, Science, Economy...). Latour admits that with this move he effectively has 'come out' as a philosopher – albeit, as he says, an 'empirical' one (Latour 2014a; cf. Maniglier 2014 and Mol 2002; see also Berliner et al. 2013) – joining in the grand, though in his case still radically diversified, project of ontology as a matter of metaphysical settlement. Indeed, this is exactly how philosophers have received him, and not least those associated with OOO (e.g. see Harman 2009; Foster 2011). Be that as it may, and notwithstanding the fact that few STS scholars have followed him in this move, we suggest that Latour's slide from methodological disruption to metaphysical model building is not accidental: It renders explicit an ontological premise that has always lurked at the heart of much of STS literature, effectively marking its limits as a reflexive project. As per our Geertzian image in the previous chapter, with STS it is reflexivity, yes, but not necessarily all the way down.

In some ways Latour's new project is similar to some of the currents of thinking that have formed part of a much-debated 'ontological turn' within the discipline of anthropology, to which we shall now turn. As several critics and commentators have pointed out (Course 2010; Laidlaw 2012, Pedersen 2012a; Vigh & Sausdal 2014; Bessire & Bond

2014; Kohn 2015) there is little agreement, and often little clarity, as to what anthropology's turn to ontology is actually meant to be, and how it relates to other recent ontological orientations within cognate fields, including, as we have just seen, philosophy and STS. Properly speaking, we are dealing here with a loose collection of ideas, arguments and approaches that appear mainly to have in common a strongly polemical tone – whatever it is, everyone who subscribes to it, as well as many of its critics, seem to think the turn to ontology is tremendously important. Still, with reference to our foregoing distinction between 'ontology' as a substantive metaphysical construction-project and 'the ontological' as a methodological orientation for anthropological analysis, it is possible to distinguish two broad tendencies,[4] and contrast them with the line of thinking that we seek to delineate in considerably more detail in this book.

The first tendency is akin to Latour's 'empirical philosophy' as it seeks to extract from the anthropological exposure to ethnographic materials elements with which to build original ontological frameworks, often with a view to providing an alternative to what is taken as the 'modern' ontological status quo. We call this the tendency towards 'alternative ontology' (we use the singular to indicate the normative propensities of this approach: modern ontology is flawed, and by looking ethnographically

[4] These two tendencies do not encompass all of the ways in which anthropologists have invoked questions of ontology in recent years. For example, in France, Albert Piette has been developing an 'existential anthropology' focused on ontological or 'ontographic' questions (2012; for alternative senses of that term see Holbraad 2003; 2009; 2012; Bogost 2012), understood as an attempt to observe and describe the entities present in any given ethnographic situation, focusing on "what *really* exists, beyond what people do or say" (Piette 2015: 97; original emphasis). Piette contrasts his 'realist' approach, as he calls it, with the writings on the ontological turn (including many of those we are reviewing here), which he sees as centring on anthropologists' 'pet themes', namely, 'differences in culture, language and relations' (ibid.: 98). Piette is correct: the approaches that are the focus of our review have in common the fact that they appeal to the notion of ontology in the context of dealing with anthropology's standing concern with apportioning similarity and difference across the diverse phenomena in which anthropologists are characteristically interested. Perhaps the most concerted account of how questions of sameness and difference take on ontological characteristics in anthropological meaning making is provided by Vassos Argyrou (2002).

45

at other lifeways, we can arrive at a better one). The second tendency inverts the priority by using a focus on ontological principles as a way of unpacking the specificities of diverse ethnographic materials at what is taken to be their deepest level. This we call the tendency towards 'deep ontologies' (here the use of the plural indicates the pluralistic character of this approach: ontology needs to be pluralized in line with the demonstrable diversity of people's lifeways). With the proviso that our intention with this distinction is not to divide anthropologists writing on ontology into clearly discernible camps, much less to offer a comprehensive overview, let us now discuss some indicative and influential contributions to these two tendencies (for other recent overviews, see Salmond 2014; Kohn 2015).

Alternative Ontology

Perhaps the most systematic and deliberate exponent of the idea that anthropological expertise can be mobilized to articulate a cogent 'alternative ontology' has been Terence Evens. Drawing on his own ethnographic engagement with kibbutz in Israel as well as a long-term project of reanalysis of Evans-Pritchard's Nuer and Zande materials, since the 1980s Evens has been providing meticulously argued critiques of concepts such as rationality, causation, logic, and other mainstays of philosophical debate, effectively using the alterity of ethnographic materials as a standpoint from which to supplant Western philosophical presuppositions, including ontological ones (e.g. Evens 1983). In his *Anthropology as Ethics: Nondualism and the Conduct of Sacrifice* (2008), arguably the culminating synthesis of this work, Evens casts his intellectual project as one of 'ontological conversion' (ibid.: xi), which he introduces like this:

What makes the following study anthropologically novel as well as radical ... is its explicitly ontological charge. Indeed, this charge recasts the discipline, not simply because it opens it to question anthropology's deepest philosophical presuppositions and directly draws inspiration from certain philosophical literature, but because at the same time it ... derives from straightforward, empirical anthropological deliberations, thus making of our discipline a co-equal partner in a philosophically received enterprise.... The claim is that the most fundamental problems of anthropological research may well yield to inquiry, but not simply by virtue of empirical analysis, however vital and necessary such analysis is. At bottom, such problems want explicit ontological deliberation. Such defining ethnographic problems as what is the nature of kinship? or how can there be order in society without government? or ... what is the sense of magico-religious presumption? are problems of otherness, and they require for their resolution nothing less radical than ontological conversion. (2008: 3)

Evens's abiding emphasis on what he calls 'ontological reflexivity' (2008: xv) in the face of the defining problems of anthropology shows how deeply germane his approach is to the line of thinking we seek to delineate in this book (Scott 2014; da Col 2014; Salmond 2014). Indeed, were it not perhaps for its 'parallel evolution' in the United States (though deeply rooted also in Evens's Manchester School training – Evens 2008: xx; cf. Evens & Handelman 2006), one could imagine Evens's project exerting a more direct, even mutual, influence on the turn to ontology we are exploring in this book. Still, as indicated perhaps by the absence of any substantive engagement between the two parallel lines of thinking (though see Holbraad 2012: xvi), there is also a basic difference between Evens's work and the ontological turn as we understand it. In contrast to the idea, outlined earlier, that ontological reflexivity is an end in itself because it constitutes a prime ingredient of the very act of anthropological description, for Evens this kind of reflexivity operates much more as a means. In particular, it serves as the major methodological premise for Evens's larger argument in favour of an ontological conversion from what he brands broadly 'dualism' (self/other, subject/object, mind/body, ideal/real etc.) to his proposal of 'nondualism', defined as a 'basic ambiguity of between-ness, an ontologically dynamic state in which boundaries connect as they separate and a thing is always also other than what it is' (Evens 2008: xx). Evens's corpus of work, then, is devoted largely to showing how such an ontological shift is necessary, not only to solve otherwise intractable anthropological problems, such as the classic debate about rationality, but also because 'it captures an experiential side of our existence that science cannot acknowledge' (2008: xi). Indeed, as the title of his magnum opus would indicate, for Evens such an approach ultimately also offers an 'ethical advantage', inasmuch as nondualism 'revitalizes the ameliorative and irenic forces of ethics [such that] the world is projected as basically and truly enchanted' (2008).

Evens's deeply Lévy-Bruhlian idea of nondualism closely parallels certain conceptual moves made by Wagner, Strathern and Viveiros de Castro (see also Scott 2013b). Moreover, the surely not unintentionally soteriological sonorities of Evens's call to 'ontological conversion' – the idea that nondualism is just morally better than dualism (see also Barad 2007; Pickering & Kuzik 2008) – might also seem to resonate with some of the political consequences of this book's version of the ontological turn, as we shall discuss in our Conclusion. Still, a key divergence may be said to lie in Evens's overriding interest in the 'we' – the first person plural of the foregoing citation. His interest in ontology is founded on a desire to arrive at a better characterization of what it is to be human – 'what makes human being tick', as he puts it in the opening sentence of his book (2008: ix). If for him, too, ethnographic contingency is a conduit for conceptual experimentation, as per our own position, that is only because it is a cipher of a deeper ontological reality which is itself universal, ethically superior and, in that double sense, necessary. Conceptual innovation in the service of an ontological regime-change, if you like. Notwithstanding his affinities with the ontological turn as we understand it, then, Evens is perhaps best considered as a particularly ontologically attuned exponent of a more diffuse strategy within anthropology, which has been branded by another of its most distinguished proponents as 'philosophy with the people in' (Ingold 1992: 696).

Indeed, Tim Ingold too is often mentioned in debates about the ontological turn and related discussions (e.g. Knudsen 1998; Henare et al. 2007; Candea 2010c), mainly for his own recommendation of a wholesale revision of not just anthropology but all disciplines, in order to rectify, as he puts it boldly, that 'single, underlying fault upon which the entire edifice of Western thought and science has been built – namely that which separates the "two worlds" of humanity and nature' (2000: 1). Drawing on an extensive and diverse range of readings, including phenomenologists such as Merleau-Ponty and Heidegger, and heterodox biological and psychological concepts such von Uexküll's *umwelt* (1934) and Gibson's affordances (1979), Ingold has spent several decades formulating a 'dwelling perspective' based on

a conception of the human being not as a composite entity made up of separable but complementary parts, such as body, mind and culture, but rather as a singular locus of creative growth within a continually unfolding field of relationships … I call this the 'dwelling perspective'. Humans, I argue, are brought into existence as organism-persons within a world that is inhabited by beings of manifold kinds, both human and nonhumans. Therefore relations among humans, which we are accustomed to calling 'social', are but a sub-set of ecological relations. (2000: 2–3)

Often in polemical style, Ingold has over the years expressed serious reservations towards a number of the ontology-oriented approaches discussed in this book (2008, 2014), just as, conversely, scholars associated with the turn to ontology have objected to what they consider Ingold's over-reliance on phenomenology (Willerslev 2007; Pedersen 2014) – an argument to which we shall return in Chapter 5. Still, his relentless and sophisticated deconstruction of the representationalist ('Cartesian') dogmas behind mainstream anthropological (and archaeological, psychological, biological etc.) thought has played a key role in the scattered topical and theoretical debates (about animism, for example; see Chapter 4) that eventually coagulated in the ontological turn.

A more recent attempt to lay the theoretical ground for an alternative anthropological ontology can be found in the work of Eduardo Kohn (2008; 2013; 2015). Like Ingold, Kohn's ambition is to formulate an 'anthropology that goes beyond the human' (2013) by 'situat[ing] all-too-human worlds within a larger series of processes and relationships that exceed the human' (2008: 6). Also very much like Ingold, he wants to accomplish this bold task by taking issue with the anthropocentric and 'linguocentric' (2008: 5) focus on 'symbolic reference' (2008: 6) that, in his view, has dominated anthropology and cognate disciplines for so long. However, whereas Ingold finds the ammunition for his attack on representationalism in different anti-representationalist scholars, such as Heidegger and more recently Deleuze, Kohn's strategy is to extend the meaning and purchase of the concept of representation itself. While he also draws on alternative biological theories marginalized by Neo-Darwinism, including von Uexküll, Kohn's main theoretical inspiration is Charles Peirce, father of semiotic theory. With explicit reference to Peirce and his tripartite distinction between indexical, iconic and symbolic sign processes (1931–1935), Kohn advances a 'posthumanist' model of representation and communication not just among but also between different life forms:

We humans live in a world that is not only built according to how we perceive it and the actions those perceptions inform. Our world is also defined by how we get caught up in the interpretive worlds, the multiple natures ... of the other kinds of beings with whom we relate... Rather than turning to ontology as a way of sidestepping the problems with representation, I think it is more fruitful to critique our assumptions about representation (and, hence, epistemology) through a semiotic framework that goes beyond the symbolic. If we see semiosis as neither disembodied (like the Saussurean sign) nor restricted to the human nor necessarily circumscribed by the self-referential properties of symbolic systems that, in any event, are never hermetic, then the epistemology–ontology binary ... breaks down. Humans are not the only knowers, and knowing (i.e., intention and representation) exists in the world as an other than human, embodied phenomenon that has tangible effects. (2008: 17)

As he has also emphasized himself (2014), Kohn's semiotic perspective differs fundamentally from the take on ontology discussed in this book. Still his project resonates closely with our central concerns, notably due to a shared emphasis on taking seriously and experimenting analytically with concepts derived from local peoples' matters of concern (about 'How dogs dream', say, 2008).[5] As he writes, 'I define ontological anthropology as the nonreductive ethnographic exploration of realities that are not necessarily socially constructed in ways that allow us to do conceptual work with them' (2015: 315). In line with our own ontological turn's aspirations, then, Kohn's project is reflexive and experimental as opposed to dogmatic and essentialist, because it blurs if not collapses conventional separations between epistemological and ontological questions (see also Pedersen 2012a).

Notwithstanding these overlaps, a noteworthy difference between Kohn's alternative ontology approach and the one outlined in this book has to do with the very direct and committed (as opposed to, in our own case, more indirect and guarded) sense in which his project is part and parcel of recent debates about the Anthropocene in anthropology (e.g. Swanson et al. 2015; Howe & Pandian 2016). For Kohn, humans have irreparably damaged life on Planet Earth, which is why 'the political problems we face today in the Anthropocene can no longer be understood only in human terms ... [but] demand another kind of ethical practice' (2014). As such, Kohn's project also speaks to wider epistemological and political agendas associated with the interdisciplinary field of interspecies ethnography spearheaded by Donna

[5] As he writes, by 'being ethnographic, and by developing conceptual resources out of this engagement, ontological anthropology, ... makes a unique contribution to what could otherwise seem to be a topic best reserved for philosophy ... Ontological anthropology is not generically about "the world," and it never fully leaves humans behind' (2015: 313).

Haraway (1988, 1991), Anna Tsing (2005, 2013, 2015), Steven Helmreich
(2009, 2012), Matei Candea (2010a, 2011a, 2014) and others whose
work can be described, to borrow an expression from Strathern (2004),
as 'partially connected' to our own ontological turn.

   This brings us to our final illustration of alternative-ontology-anthro-
pology, namely the writings of Arturo Escobar (2007), Marisol de la
Cadena (2010, 2014, 2015), Mario Blaser (2009, 2010, 2013) and more
junior scholars associated with them (e.g. Lyons 2014). Under the aegis of
what Blaser calls 'political ontology' and de le Cadena 'cosmopolitics' (cf.
Stengers 2010), these scholars have over recent years formulated a powerful, engaged and highly influential theoretical intervention in political
and political-economic anthropological debates that have for long been
dominated by various combinations of neo-Marxist and/or Foucaultian
cultural critique (e.g. Ferguson 1994; Geschiere 1997; Comaroff &
Comaroff 1998; for other critiques of this approach, see Pedersen 2011;
Bertelsen 2014; Bubandt 2014). The crux, as well as the spirit, of this
far-reaching critique is captured nicely in the following passage from
a much-cited paper by de la Cadena (2010), tellingly titled, 'Indigenous
Cosmopolitics in the Andes: Conceptual Reflections beyond "Politics" ':

Nonrepresentational, affective interactions with other-than-humans continued all over the world, also in the Andes. The current appearance of
Andean indigeneity – the presence of earth-beings demanding a place in
politics – may imply the insurgence of these proscribed practices disput-
ing the monopoly of science to define 'Nature' and, thus, provincializing its
alleged universal ontology as specific to the West: one world (even if perhaps
the most powerful one) in a pluriverse. This appearance of indigeneities may
inaugurate a different politics, plural not because they are enacted by bodies marked by gender, race, ethnicity, or sexuality demanding rights, or by
environmentalists representing nature, but because they bring earth-beings
to the political, and force into visibility the antagonism that proscribed their
worlds. (2010: 346)

Drawing extensively on Viveiros de Castro and Marilyn Strathern, as well as Isabelle Stengers (2010), Latour and Elizabeth Povinelli (e.g. 2014), de la Cadena's analytical project in several respects comes close to the one on which we focus in this book. In particular, her project shares with our ontological turn (and with Strathern's work, see Chapter 3) the notion that a 'slowing down [of] reasoning [is] ethnographically called for' (de la Cadena 2010) in order to 'take seriously (perhaps literally) the presence in politics of those actors, which, being other than human, the dominant disciplines assigned to either the sphere of nature (where they were to be known by science) or to metaphysical and symbolic fields of knowledge' (de la Cadena 2010).

Nevertheless, there are also differences between de la Cadena's approach – along with that of her sometimes comrade-in-arms, Blaser – and the one we are seeking to delineate in this book, and this in two ways: one pertaining to what is studied and another to how this is studied. For one thing, de la Cadena and Blaser's work may be described as potentially narrower in its ethnographic scope, although this may admittedly simply reflect the fact that they both happen to be working with various Amerindian peoples (much like, say, Viveiros de Castro does). Still, in reading their work, one does tend to get the impression that, to de la Cadena and Blaser, the subject matter of ontological anthropology is primarily, or even exclusively, the cosmopolitics/political ontology of indigenous peoples in different parts of the world (as opposed to the predominantly methodological focus on ontological-cum-epistemological questions that concern us in this book, as discussed in the Introduction). Connected to this is a second and perhaps more principled difference, namely that when reading these authors it is hard not to be left with the impression that 'ontology' is deployed as an attempt to 'actualize some possibilities and not others', as Blaser himself puts it (2014). Particularly in Blaser's case, this potentially normative dimension of the ontological turn, which in his own case is deliberate and made very explcit, is recognized and defended as an epistemologically, ethically and politically preferable alternative to more politically open-ended and thus (in his words) more 'risky' and even 'irresponsible' approaches (2014).

We shall have the opportunity to return to these questions in the Conclusion to this book. For the time being, and in direct continuation of the normative dimension to his work identified earlier, suffice to say that the question remains whether stipulating that doing the politics of ontology invariably means being 'ontologically political' (2014) fully lets Blaser and likeminded 'political ontologists' (2013) off the hook. Notwithstanding the caveats about their 'enacted' and 'performative' character, which do lend political ontologies a fluid and provisional nature, such arguments still operate by grounding the possibility of political difference in a prior story of how the world(s) must work, namely, in this case, the world(s) as a terrain in which ontological differences take the form of ever-emerging, fluid and tentative 'stories'. Sure, as Blaser emphasizes, the merit of this way of parsing difference is that in instantiating (enacting, performing) the very understanding of ontology it proposes this argument must itself be fluid and tentative. Nevertheless, at least for the time being and for as long as it lasts, the argument seems to cut against itself. How is the possibility of *different* differences (differences proper, we might say) not cancelled by Blaser's prior story of what differences must look like – i.e. the image of ever-emergent 'worldings', enacted and performed fluidly, tentatively and so on? The fact that in many of its elements the story itself is not altogether unfamiliar may lend some weight to this worry about the capacity of Blaser's political ontologies to differ in full enough accordance with the vast variety of ethnographic contingencies and the correspondingly varied demands they might make on anthropological conceptualization.

Deep Ontologies

The second anthropological tendency we wish to examine, which we have branded 'deep ontologies', is one that very much embraces ethnographic contingency. Rather than delineating a truer or otherwise better ontological order, here the language of ontology serves as a manner of articulating the divergent principles that may underlie the contingent diversity that ethnographic descriptions detect. Reiterating that this tendency includes diverse and often only loosely related approaches, not all of which are explicitly associated with debates about an ontological turn, it is nevertheless noteworthy that some of the main authors we think display this tendency trace their interest in ontology back to Claude Lévi-Strauss's notion of 'deep structure' (Lévi-Strauss 1963) – hence our gloss of 'deep ontologies'.[6] One prominent version of this genealogy is associated particularly with the University of Chicago and traces the link back to Lévi-Strauss via the influence he had on Marshall Sahlins and Valerio Valeri there. While neither of them lays great store by the concept of ontology as such, both follow Lévi-Strauss in paying particular attention to the categorial presuppositions that underlie particular cultural practices, which often receive their most explicit and tractable expression in cosmology, ritual and myth (e.g. Lévi-Strauss 1964; 1969; Sahlins 1985; Valeri 1995; see also Hallowell 1960; Eliade 1991). Often expressed in Lévi-Straussian oppositions between the continuous and the discontinuous, the one and the many, the complete and the incomplete (a language Lévi-Strauss himself referred to as 'qualitative mathematics' – 1954), these deep-seated assumptions about basic categories and the relations between them are then seen to

---

[6] We thank Michael Scott for emphasizing this common denominator, and for providing his own, insider's overview of the development of this line of thinking (Scott, pers. comm.). Our more brief account here is informed by Scott's, although responsibility for any errors of interpretation is of course entirely our own.

55

inform not only cosmological orders, but also historically contingent social practices and transformations. For example, in a famous study of Polynesian cosmogonic myths, Sahlins writes about the role of descent in the derivation of different cosmological elements:

> The system ... is a veritable ontology, having to do with commonalities and differentiations of substance. Relations logically constructed from it – e.g. heavens are to earth as chiefs are to people – are expressions of the essence of things. Hence the relations and deeds of primordial concepts as represented in myth become, for the persons descended of such concepts, the paradigms of their own historical actions. (Sahlins 1985: 14)

It would be wrong, of course, to attribute attention to the deep ontological structures and dynamics informing different sociocultural orders and transformations to Sahlins and his Chicago colleagues alone. For example, Bruce Kapferer's study of what he calls the 'political ontology' of Sinhalese statecraft has arguably been just as influential (Kapferer 2010; Taylor 2001). A little as with Sahlins's argument about Polynesian descent, however, the Chicago line is noteworthy for the manner in which the deep-structural principles of indigenous ontology formulated by Sahlins and Valeri have become paradigms for the (anthropological) actions of a number of Sahlins's most distinguished students, including Gregory Schrempp (1992; 2012), Michael Puett (2014), and Michael Scott (2007).

Scott in particular has been the most deliberate in developing this line of thinking into an anthropological program of 'comparative ontology', as he calls it. Broadening out from Sahlins's emphasis on cosmogony (though see Scott 2015b), he proposes the notion of 'onto-praxis' as a manner of elucidating the 'root assumptions operative [in a given historical or social context] concerning the essential nature of things and their relationships within multiplex, and at times even contradictory, cosmological schemes' (Scott 2007: 3). Crucially, these root assumptions run deeper than the kinds of concepts or phenomena with which anthropologists habitually frame their analyses. So, anthropological rubrics such as 'power, knowledge, identity, hybridity, etc.', as well as ethnographically isolated socio-cultural phenomena such as 'land tenure, land disputes, leadership, violence, etc.', are better understood as contextual manifestations of 'specific mappings of the number, nature, and interconnections among fundamental categories of being' (2007: 3–4). In his own study among the Arosi on the island of Makira in the Solomon Islands, for example, Scott shows systematically how the postcolonial transformations of such ethnographic domains as kinship, political leadership, land tenure, religious worship and, most importantly, the structure of exogamous landholding matrilineages that are deemed to be exemplarily autochthonous to the island manifest a set of root assumptions he brands as 'polyontology'. This he defines as an 'understand[ing] of the universe as the sum of multiple spontaneously generated and essentially different categories of being' (2007: 12), the most clear evidence of which, as he explains, is found in Arosi 'narratives of the independent autochthonous origins ... of each matrilineage' (2007). In such a poly-ontological situation, the constitution of society depends on creating 'unifying relations among multiple pre-existing categories of being' (2007: 18), and hence the importance of local practices of exogamy, sharing of land, and hospitality (2007: 12; 163–260). Drawing on philosopher Roy Bhaskar's notion of 'ontological stratification' (Bhaskar 1994; see also Graeber 2015), Scott articulates the relationship between root ontological assumptions and the social practices that manifest them in terms of a model of relative depth:

[T]he primordial condition of originally separate and autonomous groups is understood to be the permanent foundational ontology on which a secondary structure of relations among disparate groups has been socially constructed... Although each level of reality in a stratified ontology entails practical tendencies, these emerge from, and are influenced by, the deepest level of being. (ibid.: 13)

It is relevant for our purposes to note that one of Scott's main concerns in developing this model is to counteract what he sees as the undue dominance in Melanesian anthropology of the so-called Melanesian Model of Sociality, which he ascribes particularly to Roy Wagner and Marilyn Strathern, two of the prime movers of the line of thinking with which this book is concerned. Strictly inverse to his, Scott argues, this model is 'mono-ontological': it 'posits the primordial oneness of all things' (Scott 2007: 18), as expressed in the relational flow of Melanesian sociality (Wagner 1977b; Strathern 1988), which lends personhood its 'partible' and 'fractal' character (e.g. Strathern 2004; Wagner 1991), such that 'the burden on [social] praxis is to achieve and maintain differentiation (Scott 2007: 18). To the extent that this ontological monism is now a 'nascent orthodoxy' (ibid.: 31) in Melanesian anthropology, Scott suggests, the Arosi study offers a counterexample that may also provoke us to re-examine its applicability in different parts of the region, including even the Highlands of Papua New Guinea, where Wagner and Strathern conducted fieldwork (ibid.: 24–32). We shall return to this debate in Chapter 6.

In more recent writings, Scott has extended his critique of the dominance of particular ontological models in anthropology by examining how they format also certain strands of anthropological theory (Scott 2005; 2013a; 2013b; 2014). Treated ethnographically as contingent configurations of thinking and acting, theoretical trends in anthropology too manifest 'root assumptions concerning the essential nature of things and their relationships', as per our earlier citation of Scott on onto-praxis. In fact, a prime target of Scott's critiques has been the literature on the ontological turn itself. He suggests that at the root of diverse writings on this theme, including not only by Wagner and Strathern, but also Viveiros de Castro, Ingold, Latour, Evens, Kohn, Elizabeth Povinelli (2002), as well as the two of us, is the ontological assumption of what he calls 'nondualism', citing Evens (2008). Presented as the inverse of Cartesian dualism (a 'poly-ontolog[y] of the simplest kind [that posits] a plurality of essential qualities, but in this case only two' – Scott 2014: 34), nondualism emphasizes 'the perpetual movement and mutability of that which is always already mixed – a kind of ceaselessly surrealist metamorphosis or recycling of transiently stable and co-mingling beings' (ibid.: 35). Indeed, Scott is careful to note that the nondualist ontology that he thinks is indigenous to much of contemporary anthropology comes in two variants. For some, 'the nondualist orientation of their ethnographic consultants constitutes a fuller apprehension of the true flow and ambiguity of being and becoming' (here he cites as examples Evens and Ingold among others). For others, nondualism is to be elevated as a methodological principle that allows anthropologists to 'adopt a position of apositionality, a motile analytical transit that, because it is *potentially* every theoretical position, everywhere and every-when, is simultaneously no theoretical position, nowhere and no-when' (ibid.: 37) – this being an excellent gloss on the kinds of positions and methods to which the rest of this book is devoted. Indeed, the distinction coincides with the one we are making here, between the 'alternative ontology' tendency and the methodological orientation we are seeking to delineate in this book.

We shall have occasion later in the book to address the important idea, which Scott is not alone in expressing (see also Heywood 2012), that the ontological turn, including in the version we present here, manifests a particular 'meta-ontology' of its own. But regardless of the question of how far this claim is born out (we shall see that to an extent it is, though perhaps not as absolutely as Scott would have it), we note here that Scott's profoundly reflexive move of applying his onto-practical analytical framework to the practice of anthropological theory itself closely parallels the kind of reflexivity we seek to elucidate in this book. Similarly, although it is beyond the scope of the present discussion to assess the rights and wrongs of Scott's critique of mono-ontology in Melanesianist anthropology, we note that Scott's insistence on the difference that his ethnography of the Arosi can make to this debate is very much in line with the idea, central to this book, that the experimental conceptualizations that are so central to the ontological turn as we understand it draw their power from just such ethnographic contingencies. Indeed, it is just because of its embrace of ethnographic contingency – the idea that ontological positions must be pluralized in response to the varieties of ethnographic experience – that the 'deep ontologies' tendency that Scott exemplifies comes closest to the approach we are seeking to elucidate in this book.

Still, there is also a significant, though by no means absolute, distinction to be made between the 'deep ontologies' tendency and the methodological turn to ontology we are outlining. This comes down to the core idea that ontologies are to be located 'deep within' ethnographic materials, the job of the anthropologist being to bring them to light by analysing their 'manifestations' on the surface of social praxis. Perfectly captured in Scott's Bhaskarian image of ontological strata, we would note that this is itself a very particular way of formatting the object of anthropological inquiry. So, backhandedly to return the compliment of Scott's reflexivity, we may query the meta-ontological assumptions of the 'deep ontologies' approach, and particularly the idea that the human world comprises a deep level of ontological structures and dynamics which in turn informs a shallower ontological level of social formations and practices. For one thing, one can hardly fail to notice the historical and intellectual contingency of this image, recalling the sheer gravity of such notions as *on*, *Sein* or Being in particular traditions of the Western philosophical canon. What is more, this way of 'locating' ontologies, namely finding them deep within geographically and historically distributed ethnographic situations, reiterates, albeit at a different ontological 'level', more standard ways of thinking about, say, cultures. If ontologies, like cultures, are objects to be found out there in the world (and note the ontological weighting of the idea), then the spectre of all the discomfiting questioning to which the notion of culture has been subjected in the past half-century returns: where does one 'indigenous ontology' end and another begin, how are we to theorize the interactions between them, what about change over time, how to avoid the risk of reification and

Orientalism, and so forth (see also Venkatesan et al. 2010; Alberti et al. 2011; Keane 2013).[7]

But from the point of view of our line of thinking here, the biggest question mark has to do with the limits that the 'deep ontologies' tendency places on the possibility of conceptual experimentation. Building into its own premises particular ways of conceiving not only of ontology, but also of society, praxis, cosmology and other such mainstays of anthropological reflection and debate, it would seem that its tendency is to exclude them from the scope of the reflexive conceptual versatility that we are seeking to propound here. For example, Scott's proposal that social forms and practices rest on underlying ontological assumptions invokes a series of anthropological assumptions about what might count as a social form or practice in the first place. Left largely unproblematized, such assumptions are ratified in his analysis of the Arosi material, as, for example, when he presents the tension between the underlying poly-ontology and practices of leadership which precariously promote 'social cohesion' between ontologically diverse matrilineages. But faced with the thoroughly anti-social (or at least anti-Durkheimian) nature of poly-ontology as Scott characterizes it – he defines poly-ontology also as 'a cosmos in which the parts precede the whole' (Scott 2007: 10) –, one may fairly wonder what might even count as 'social cohesion' in such a cosmos. Probing such a question would at the very least involve shifting the conceptualization of society away from the standard Durkheimian image of a whole that precedes its parts. There is no reason to assume that the deep ontologies tendency outright precludes this kind of reflexive experimentation with the conceptual infrastructure of anthropological thinking, i.e. the turn of thinking to which this book is devoted. But the meta-ontological weight of this way of locating ontologies, deep in

---

[7] Much of the energy of writers in deep ontology has indeed been devoted to giving new, ontologically inflected responses to these otherwise familiar-sounding questions (e.g. Scott 2007: 229–300; see also Kapferer 2010; Lloyd 2007; 2011; 2012; cf. Luhrmann et al. 2013).

the world, roughly underneath societies and cultures, does seem to get in the way.

A similar critical perspective can be taken on an approach advanced by Philippe Descola, which has been at the centre of debates about anthropology and ontology in France, and is now so influential also among English-speaking anthropologists that for some (not least in the United States) it has become almost synonymous with the 'ontological turn' of the discipline (although see below Descola's polite disclaimer; cf. Kelly 2014; Lenclud et al. 2014; Kohn 2015). Put forward with magnificent clarity and detail in his magnum opus, *Beyond Nature and Culture* (2013), Descola's approach takes the form of a comparative ontology that is in some ways analogous to Michael Scott's proposal. While tracing a more direct line back to Lévi-Strauss, whose post he held at the Collège de France, Descola's anthropological objective too is to push anthropological analysis to the deepest level at which differences between human lifeways can be registered, which for him too is the level of ontology. As he writes:

My conviction is that systems of differences in the ways humans inhabit the world are not to be understood as by-products of institutions, economic systems, sets of values, cultural patters, worldviews, or the like; on the contrary, the latter are the outcome of more basic assumptions as to what the world contains and how the elements of this furniture are connected... My only claim in the so-called 'ontological turn' – an expression I have never used myself – is just one of conceptual hygiene: we should look for the roots of human diversity at a deeper level, where basic inferences are made about the kinds of beings the world is made of and how they relate to each other. (2014: 273)

Much as we saw with Scott, then, for Descola ontologies are plural and are to be found somewhere underneath, or 'upstream' as he puts it (2013: 115), the phenomena that ethnographers record, and can be used anthropologically to arrive at the principles that account for their diversity. The major difference from Scott and the Chicago culturalist tradition, however, is that for Descola these ontologies are not to be read off the surface of ethnographic phenomena inductively (i.e. asking, as with Scott, what the underlying ontological assumptions of a given ethnographic configuration are), but are rather established deductively in a near-transcendental manner:

[O]ntological discrimination does not stem from empirical judgements regarding the nature of the objects that constantly present themselves to our perception. Rather it should be seen as what Husserl called a prepredicative experience, in that it modulates the general awareness that I may have of the existence of the 'other'. This awareness is formed simply from my own resources – that is to say, my body and my intentionality – when I set aside the world and all that it means for me. So one could say that this is an experience of thought prompted by an abstract subject. (2014)

From such an abstract cogitation of the ego, then, Descola proceeds to deduce his famous scheme of four distinct ontologies, in a manner entirely reminiscent of a structuralist parsing of binary oppositions. Careful to adduce diverse ethnological evidence showing that the aforementioned distinction between body and intentionality, branded more technically as a distinction between 'physicality' and 'interiority', is not merely an ethnocentric projection of Western dualism but rather an elementary contrast present in all known cultures' (2014: 116–21), he uses this opposition to delineate four possible ways in which ego can relate to others:

Faced with some other entity, human or nonhuman, I can assume either that it possesses elements of physicality and interiority identical to my own, that both its interiority and its physicality are distinct from mine, that we have similar interiorities and different physicalities, or, finally, that our interiorities are different and our physicalities are analogous. I shall call the first combination 'totemism', the second 'analogism', the third 'animism', and the fourth 'naturalism'. These principles of identification define four major types of ontology, that is to say systems of the properties of exiting beings; and these serve as a point of reference for contrasting forms of cosmologies, models of social links, and theories of identity and alterity. (2014: 121)

The power of Descola's model is demonstrated not only by its all-encompassing ethnological purchase, rendering ethnographic materials from across the world at once intelligible and comparable, but also in the way it recasts decisively some of the foundational debates of the discipline (e.g. about animism, totemism, and so-called apparently irrational beliefs in general). Furthermore, in relation to our concern here with the critical potential of the turn to ontology, Descola's model arguably achieves a thoroughgoing and principled 'provincialization' (*sensu* Chakrabarty 2000) of Western ontological assumptions, relegating them to just one out of four possible ontological permutations. Still, by the same token, and in line with the deductive character of the model, we may note that this pluralization of ontology is by no means meant as a wholehearted embrace of ethnographic contingency. Far from it:

[The patent imperfection of scholarly models such as Descola's own stems] from the fact that they are unable to take into account the infinite richness of local variants. But that is the risk run by any attempt to generalize, which has to sacrifice the spicy unpredictability and the inventive proliferations of day-to-day situations in order to reach a higher level of intelligibility regarding the mainsprings of human behaviour. (2014: 115)

Putting matters in terms of our argument about the relationship between reflexivity and conceptualization in the ontological turn, we might say that Descola's attempt rigorously to reconceptualize the ontological foundations of human behaviour is characterized by an overt desire *not* to do so reflexively, i.e. by adumbrating the contingencies of ethnographic materials in the manner that we have been discussing here. Rigour, for him, is a matter of subjecting such contingencies to the organizing powers – indeed, the deductive necessity – of anthropological thought. Needless to say, doing this requires a strong dose of ontological 'foundationalism', establishing the terra firma of basic ontological premises upon which to found the edifice of anthropological theorization – tellingly and vividly, Descola writes of his argument as an 'architectural plan for a new communal house that would be more accommodating to nonmodern cosmologies' (2014: xvi-xvii). The entirely deliberate and principled manner in which Descola goes about the task, as we have seen, renders his approach a perfect inversion of the one we seek to present in this book: an 'ontological turn', if such it is, constitutively disinterested in turning on itself (see also Latour 2009). No wonder, then, that Descola is reluctant even to use the expression.

Our Ontological Turn

In this chapter, we have attempted to draw clear, principled distinctions between four 'other' ontological turns and the version this book is about. As an exercise in broad intellectual heuristics, we have sought to map out some of the general paths of thinking that can be detected in the literature on the ontological turn, the better to set up the coordinates of the exposition to which the following chapters are devoted. We should make very clear here that our tendency in this chapter towards 'purification' (if not quite in Latour's 'modern' sense) is the by-product of our attempt to present in terms that are consistent and clear bodies of writing whose subtleties inevitably go beyond the purposes of our discussion here. So, as we have tried to indicate without going into the full exegetical detail, particular authors that we have identified with one line of thinking may often express thoughts or make arguments that are more in line with another, or indeed with the perspective we seek to articulate in this book.

For instance, notwithstanding their abiding argument in favour of nondualism as a superior 'alternative ontology', Evens's methodological reflexivity and Ingold's critique of representationalism have provided some of the most distinguished examples of the kind of conceptual experimentation we here associate with 'our' ontological turn – an observation that may be extended to the work of Kohn, De la Cadena and Blaser.

Similarly, while we would associate Kapferer's classical works on political ontology with the tendency to look for 'deep ontologies', more recently Kapferer has been making concerted arguments to the effect that the radical intellectual potential of anthropology depends on the virtue that it makes of the 'exotic' – the 'outside of thought', which is sharply to be distinguished from the tendency to exoticism (2013). Furthermore, some of the research conducted at the increasingly creative interface between anthropology and STS, by scholars such as Casper Bruun Jensen, Brit Winthereik, Marianne Lien, Antonia Walford, Hannah Knox, Atsuro Morita, Astrid Andersen, Penny Harvey, Steven Helmreich and others, is both methodologically and theoretically closely aligned with the approach that concerns us here, showing how its purchase can extend deep into contemporary debates about such topics as technology, infrastructure, 'big data', and so on – topics that may seem far removed from the concerns of figures such as Roy Wagner, Marilyn Strathern and Viveiros de Castro, whose work we shall be exploring in detail.

Indeed, while in the following chapters we present these three authors as forming the core intellectual trajectory of 'our' (and by the token of our discussion 'their') version of the ontological turn, we should make clear here that they are by no means consistently fixed on, or by, this manner of thinking either. For example, the objectifying (and, as we saw, typically foundationalist) idea that different 'ontologies' can in some sense or other be ascribed to different groups of people, or to different ethnographic practices, does sometimes appear even in the writings we are about to review in the following chapters (e.g. Pedersen 2001; Holbraad 2012: 78; Viveiros de Castro 2012). And, as we shall see in some detail in Chapter 4, so does the notion of 'multiple ontologies', sometimes conveyed through the potentially reifying (and essentializing) image of 'multiple worlds' (e.g. Henare et al. 2007: 12; cf. Alberti et al. 2011).

By the same token, as we shall see particularly in relation to the work of Wagner and Strathern, some of the insights and turns of argument that we present as part of the ontological turn's trajectory make little reference to the notion of ontology at all. We shall argue, nevertheless, that these writings consistently lend themselves to a synthetic exposition of the core orientation that this book seeks to articulate. So our task in presenting this line of thinking, showing its pitfalls as well as its strengths, is neither exegetical nor an exercise in intellectual history in the tradition of George Stocking (e.g. 1987, 1995) and other historians of anthropology. Indeed, while in the chapters that follow we shall go deep into the writings of the three key figures we have selected, we do so not in order to present a judicious review of their writings, but rather in order to extract from them the line of thinking we want to expose. Our aim, in other words, is not to be comprehensive or encompassing in our account of the works we present, but rather to convey as clearly as we can certain core moves of argument that they contain. Our task is interpretatively to mould these arguments into an intellectual exposition, showing how it runs through the work of the authors we review.

The rest of this book, then, traces the development of a certain form of anthropological thinking and mode of ethnographic description that has led up to the current debates about the ontological turn within anthropology. Having examined in this chapter a selection of the most influential contenders to this label, we shall now proceed to use the expression 'ontological turn' in a deliberately programmatic way, to refer only to the strand of thinking with which we are primarily concerned in this book. We use the expression advisedly, and not without hesitation, for it is fast becoming a kind of intellectual brand, at least according to some of its most ardent critics (Laidlaw 2012; Vigh & Sausdal 2014, Bessire & Bond 2014; Graeber 2015). We hasten to add that, while engaging with it critically and acknowledging that its name is not always helpful, our goal is to demonstrate that the ontological turn is far more than a passing fad.

Indeed, keeping the foregoing critical discussion of the four alternative approaches in mind, we are now in a position to state in a nutshell the most distinctive feature of the version of the ontological turn that we consider 'ours' and which we seek to delineate in the remaining chapters of this book. This, we suggest, is its abiding concern with *freeing thought* from all metaphysical foundationalism – whether substantive or methodological, normative or pluralistic. Or, if this sounds rather too ambitious, at least strategically to displace the search for ontological foundations to give precedence to a distinctly anthropological task, namely that of giving full expression to the contingencies of a given ethnographic situation. This is significant because, in principle, these contingencies can cut against not only the metaphysical commitments of the ethnographer, but also the ontological binds within which anthropological methodology is meant to operate. In other words, as we explained in the Introduction, insofar as this way of doing anthropology is concerned with matters of metaphysics or 'ontology' in the philosophical sense of the word, it is so in the service of allowing the object of ethnographic analysis to have a transformative effect on the ontological assumptions the anthropologist brings to it, and in that way contribute to setting the terms for the anthropologist's conceptualization and analysis of it. The point is not to keep looking for new alternatives to what the world is like. Rather, it is to find ways to allow the world, as it expresses itself in the contingent ethnographic situations that we encounter as anthropologists, to show us how things could be otherwise. Posing ethnography as a conduit for metaphysical contingency, anthropology turns to ontological questions without taking any single 'ontology' as an answer – its ultimate concern being not with what is, but with what *could be*. To set this argument on its tracks, let us now turn to the work of Roy Wagner who, as we shall see, essentially set it in train in the first place.

TWO

❧

Analogic Anthropology: Wagner's Inventions and Obviations

Consider the work of improvisation, say of a jazz musician. As a self-respecting craftsman, he knows his instrument inside out. He practices for hours every day, maintaining and perfecting his technique (scales up, scales down, arpeggios, trills …). All this preparation he has literally at the tips of his fingers when he goes on stage. The musical skills he's been honing for so many years are now the resource on which he draws for his performance. Listening to the other musicians, responding to their musical inventiveness, adumbrating it on his own instrument, enlarging upon it, taking it in a new direction to come up with something exciting – a solo! That's what people come to listen to, and that's what he came here to do.

What if one were to think of human social life in terms of performing solos, rather than practicing scales and arpeggios? What if we stopped assuming that the best way to understand what people get up to is to chart the 'capabilities and habits acquired by man as a member of society', to take a classic definition of the anthropological concept of 'culture' (Tylor 1920: 1)? That is to say, what if we stopped assuming that our job as anthropologists is to chart people's belief systems, language rules and practices, values, laws, customs, kinship arrangement, bodily habits and techniques, and all the other patterns of conventional behaviour that make human society look like one big music practice room? What if, instead, we did an anthropology of solos, or even anthropology *as* solos? What do people's lives look like if we think of them as attempts to subvert the established conventions, to bring about something new and fresh? And crucially, how would the anthropological notion of 'culture' need to be redeployed – reinvented or obviated even – to allow anthropologists themselves to go beyond their penchant for charting conventions, patterns of behaviour, norms and the like? Can anthropology play itself out of its *own* culture, namely that of 'culture', like a jazz soloist plays himself out of scales and arpeggios?

Roy Wagner's work constitutes the most deliberate and systematic attempt to explore this possibility for anthropology. As American as jazz and the beatniks, Wagner has used his lifelong ethnographic engagement with Melanesia – a place where, for him, scales are taken for granted and solos are what life is all about – as a point of departure for redefining the very idea of 'culture' – an idea which, notwithstanding its Germanic roots, has been itself definitive of the American tradition of 'cultural anthropology'. His project has been to reinvent cultural anthropology by inventing a new concept of culture, to wit one that proposes culture as a process of invention.

Now, Wagner does not talk much about ontology, and his experiments with the concept of culture predate the recent furore about the ontological turn in anthropology by a number of decades (although, since its advent, he has been sympathetic – e.g. see Wagner 2012a). Nevertheless, we suggest that, in its substance if not its letter, Wagner's anthropological oeuvre lays the ground for what we take to be the signature of the ontological turn, namely the willingness to stage the encounter with ethnography as an experiment in conceptual reflexivity. Indeed, as we shall see, Wagner's reinvention of the concept of culture in the light of his ethnographic encounter with Melanesian lifeways is not only a prime example of what we take the core move of the ontological turn to be, but also the first argument *for* such a move. The ontological turn is a turn towards what Wagner has in mind when he talks about culture as invention. It is, if you like, a turn to an anthropology versed in scales and geared towards undoing them in solos.

This chapter demonstrates this claim by first outlining the central tenets of Wagner's anthropological invention of culture and then exploring its consequences with reference to Wagner's work on myth in particular. Wagner's symbolic analyses of myth, for which he develops a technical vocabulary and elaborate interpretative procedure centred on the concept of 'obviation', does not only demonstrate his experimentation with the notion of culture, but also reveals particularly clearly the ontological implications of his approach. To set the stage, we begin by outlining the manner in which Wagner's invention of invention sets itself up in relation to the prevailing conventions of anthropological thinking, and particularly the North American tradition of cultural anthropology upon which Wagner sought so drastically to innovate.[1]

American Convention

We have already suggested that, in a variety of different ways, the turn to ontology in anthropology has been bound up with a critique of the distinction between nature and culture (and the kindred distinction between nature and society) as the founding anthropological matrix for distributing similarity and difference. Wagner's wholesale reinvention of the concept of culture can be seen in this light. In particular, his experiment with the concept of culture constitutes a concerted attempt to rid a particular strand of the anthropological critique of nature/culture thinking of its central contradiction, namely that it is so often done in the name of cultural relativism – the prime tenet of the Boasian tradition of American cultural anthropology. According to this view, the very

---

[1] This is not the place to go into any detail about the degree to which Wagner's project might be deemed representative of the US tradition of cultural anthropology, insofar as it makes sense to speak of such a unified intellectual tradition in the first place. But it is perhaps worth pointing out that the thinking of his PhD supervisor David Schneider, and possibly that of Wagner himself too, was framed by what has been described as the 'schism' between 'socially' and 'culturally' oriented anthropologists in the department of anthropology at Chicago; a schism

concepts of nature and culture are themselves cultural constructions – a mark of anthropologists' 'Western', 'Cartesian' mindset – and therefore should not be projected ethnocentrically onto others who may not classify the world in these particular terms. Once stated like this, the oddity of this position, which verges on self-contradiction, becomes obvious. It is only by relying on the distinction between nature and culture – precisely the form of 'classifying the world' we anthropologists are supposed not to project onto others – that we are able (indeed bound) to repudiate the 'ethnocentrism' of the distinction itself. Anthropologists should not universalize the distinction between nature and culture, we say, because other cultures do not make it, and thus we reinscribe its universality in the very act of denying it.

sometimes attributed to the employment of Radcliffe-Brown in 1931 and Boas' student Edward Sapir in 1925 (see Darnell 1986: 159–63; Kuper 1999: 128–30). Be that as it may, Schneider's idiosyncratic anthropological project, to which Wagner frequently acknowledges his debt, represented a creative combination of several intellectual traditions, including the American functionalism developed by Talcott Parsons, with whom Schneider, along with Clifford Geertz, had studied at Harvard (Kuper 1999: 69–73, 124–7). Indeed, Schneider's brand of 'symbolic anthropology' may be described as an amalgamation of Parsons, Lévi-Strauss and Sapir/Benedict, in the sense that 'symbols constitute an autonomous system; within this system, certain symbols are central points of orientation on which all else depends. The claim that culture is a system of meanings which cannot be reduced to accounts of individual behaviour derives from Parsons; the emphasis on the system as a set of relationships derives to some extent from Levi-Strauss; the focus on the cultural core or distinctive essence of an apparently complex society derives from Ruth Benedict and the later work of the culture-and-personality school on national cultures' (http://what-when-how.com/social-and-cultural-anthropology/symbolic-anthropology/; see also Buckley 1996 and especially Darnell 1986 on the crucial differences between Sapir and the other scholars associated with the 'culture and personality' school; differences which seem to point to interesting theoretical continuities between Sapir, Schneider and Wagner). Certainly, 'Schneider went beyond Parsons. Not only is the symbol arbitrary [as in Saussurean structural linguistics], but the referents, the things or the ideas for which the symbols stand, are themselves cultural constructs. They may have no objective reality at all' (Kuper 2000: 133). As we are going to see, it was just this idea – that symbols in a paradoxical yet fundamental way stand *for themselves* – that Wagner took over from his supervisor and developed into his own sophisticated theory of symbolic process.

Wagner's critique of the culture concept can be characterized as a concerted attempt to overcome this basic paradox, rescuing cultural relativism from itself. The question is this: How might the 'non-universality' of the distinction between nature and culture be conceptualized *without* recourse to that very distinction? And what might the implications of such a way of thinking be for the practice of anthropology? Although charting the varied sources of debate that have led up to this type of question is too large a task to undertake here, it is worth noting that the critique of the anthropological matrix of nature versus culture has arisen within the discipline as an extension of anthropologists' longstanding investment in what has come to be known as 'cultural critique' – a mode of anthropological argument that stems from the inclination to 'relativize' things by using ethnography to show how they could be imagined differently. Cultural critique in this inclusive sense has cut across the discipline's otherwise determining theoretical divides and national traditions (Hart 2001) – a point to which we shall return in the Conclusion.[2] Nevertheless, with its abiding concern with cultural relativity, American cultural anthropology has provided the most fertile ground for the cultivation of this critical stance, as shown, for example, by the record-breaking sales of such books as Ruth Benedict's *Patterns of Culture* (1934) and Margaret Mead's *Coming of Age in Samoa* (1961) – both stemming directly out of Franz Boas's grandly relativizing anthropological project (Boas 1940; see also Geertz 1973).

---

[2] In an incisive discussion, Keith Hart shows that the task of relativizing assumptions that other disciplines as well as the wider public may take for granted has been a major part of anthropologists' intellectual mission from the early years of the discipline. For example, the positivist outlook of Durkheimian sociology did not stop Marcel Mauss from using Maori prestations as a vantage point from which to criticize modern markets any more than the relativist premise of Boasian culturalism dictated to Margaret Mead that she should use her fieldwork among adolescent girls in Samoa to show up the peculiarities of American parenting (Mauss 1990; Mead 1961).

Analogic Anthropology

The rise of postmodernism in the 1980s in the United States, promoting a 'repatriation' of anthropology in which not only sundry aspects of modern Euro-American culture but also, crucially, anthropology itself become the object of anthropological reflection, can be seen as the ultimate logical consequence of a cultural critique that is played out from within the coordinates of cultural relativism (e.g. Marcus and Fischer 1986). Anthropology itself, including its *ur*-binary of nature versus culture, is thus relativized as one among myriad cultural constructs. Remarkably, however, what often goes unmentioned when the story of late twentieth-century cultural anthropology gets told (e.g., Kuper 2000; Barnard 2000) is that the same tradition of cultural critique that bore us the postmodernist crisis of representations in the 1980s also delivered a turn of thinking that was both less inward-looking and more radical in its critical and reflexive implications: instead of relativizing anthropology as a cultural construct, this tack of research relativized culture (and its opposition to nature) as an anthropological one.

The turning point arguably was David Schneider's *tout court* attempt to debunk the anthropological concept of kinship. Following his original research in Micronesia on Yap kinship categories, Schneider 'repatriates' his project by examining the cultural construction of kinship among middle-class residents of Chicago (1968). There he finds that the 'core symbols' of American kinship, as he called it, are organized around a culturally elaborated distinction between what are deemed to be 'facts of nature' (particularly sexual intercourse and the blood-ties to progeny it is supposed to engender) and the cultural conventions that are deemed to bring them under control (particularly the codes of conduct that are enshrined in marital law, and the kin-relationships to which they give rise). To the extent that this distinction also informs the cross-cultural study of kinship by anthropologists (from Lewis Henry Morgan's distinction between descriptive and classificatory kin terms to Lévi-Strauss's contention that kinship systems offer varying solutions to the universal problem of humans' passage from nature to culture), the very concept of kinship is an ethnocentric projection of a Western cultural model (see Schneider 1968; see also Leach 2003: 21–32).

To be sure, Schneider's argument could be viewed as a straightforward example of relativist cultural critique US-style, which is how Schneider himself presented it. What brings Schneider's critique to the brink of something more than just an argument from cultural relativism, however, is his willingness to adumbrate the implications of the cultural contingency of the nature/culture distinction for the infrastructure of anthropological analysis itself, thus prefiguring the kind of conceptual reflexivity that lies at the heart of the ontological turn. Indeed, while his critique can also be seen as a pioneering example of an anthropological strategy that has since become very familiar, namely that of weighing up the ethnocentric baggage of all manner of categories that had previously been taken for granted as supracultural analytical tools ('religion', 'ritual', 'politics', 'labour', 'property' and so on), Schneider's critique amounted to more than that by virtue of the fact that its target was a category as foundational to the discipline as kinship. As borne out by his own subsequent reflections on the matter (1995), Schneider was well aware that to undermine the natural basis for the study of kinship was to undermine willy-nilly the most basic premise of the anthropological project of cross-cultural comparison itself.

What made kinship different from religion, ritual, politics and so on, giving it its privileged position as the bedrock for cross-cultural comparisons, was that it was assumed to be the most basic point of contact between the universal facts of human nature and their variable cultural elaborations by different social groups. For Schneider, to explode the study of kinship was to deal a blow on the foundational matrix through which anthropologists conceptualized similarity and difference itself. Of course, as Adam Kuper remarks in an acerbic review of Schneider's contribution to American cultural anthropology, it is unsurprising that Schneider did not take his maverick move the whole way so as to undermine the anthropological notion of culture alongside that of nature

(Kuper 2000: 122–58). To do so would have nullified the premise of his critique, namely that kinship, and the distinction between nature and culture upon which it is based, is an American cultural construct. It would radicalize his relativism to such an extent that it would end up undoing itself.

This is exactly what Roy Wagner did in *The Invention of Culture* – a book dedicated to Schneider, who was Wagner's teacher, with an acknowledgement of the 'germinal' character of its debt to him (Wagner 1981: ix). Like Schneider, Wagner is concerned with the work that the distinction between culture and nature does for anthropology and, again like Schneider, he contrasts this distinction to ways of thinking and acting elsewhere so as to show that it exemplifies distinctively Euro-American presuppositions. But what allows Wagner to escape the paradox of charting such contrasts in terms of cultural difference is that, unlike Schneider, he made the concept of culture *itself* the target of his critique. To substantiate this crucial point, we shall now present Wagner's argument in some detail, drawing not only on *The Invention of Culture* but also on its ethnographic prequel, *Habu: The Innovation of Meaning in Daribi Religion* (1972), where Wagner developed the core elements of his model with reference to his fieldwork among the Daribi people of the Highlands of Papua New Guinea.

Culture as Invention

A number of largely tacit assumptions lie behind the American cultural anthropological conventions that Wagner set out to lay bare and criticize in *Habu* and *Invention of Culture*. In line with the word's etymological root in the Latin *colere*, 'to cultivate', Wagner argues (1981: 21), anthropologists imagine 'culture', in its broadest sense, as a set of conventions by which people order and make sense of themselves and the world around them (again, in the broadest sense, nature). Since conventions established at different times and places by different groups of people vary, and the very idea of a convention implies a particular social setting in which it is abided, as anthropologists we like to speak of 'cultures' in the plural.[3] The languages people speak, their social arrangements and political institutions, their means of subsistence, technological wherewithal, economic activities, ritual practices and religious beliefs, ways of seeing the world, perhaps even their ways of feeling in it – all these, we consider, are cultural 'conventions' that people establish and live by. So our job as anthropologists, we are trained to assume, is to describe the conventions of the people we study, having learned something of their inner workings as we ourselves lived by them during fieldwork. And since, we take it, these conventions are ultimately 'artificial' (Wagner 1981: 49), in the sense that they are established in order to organize and make sense of a world that is prior to human action and is by this token 'given' or 'innate' (ibid.), as anthropologists we are also charged with accounting for the conventions we describe. We may interpret how and why the conventions in question make sense to the people who live by them, explain how and why they emerged as they did, and even draw conclusions about humanity in more general terms – in short, we may engage in the project of anthropological analysis and theorization in its varied and competing guises.

Much as with Schneider on kinship, Wagner's critical move in all of his main early works is to relativize this set of assumptions by showing their peculiarly 'modern Western' character, as he calls it (e.g. 1981: 7), and using ethnography from elsewhere to explore alternatives. On the former count, he observes that the assumption that human activity is directed towards gaining a handle on the vagaries of an otherwise disorderly and

---

[3] As Wagner explains: 'When [anthropologists] speak as if there were only one culture, as in "human culture," this refers very broadly to the phenomenon of man; otherwise, in speaking of "a culture" or "the cultures of Africa," the reference is to specific historical and geographical traditions, specific cases of the phenomenon of man. Thus culture has become a way of talking about man, and about particular instances of man, when viewed from a certain perspective' (1981: 1).

unpredictable world, or 'nature', by establishing shared cultural conventions lies at the heart of the way we think of sundry aspects of life in Western societies. Schneider's own depiction of the regulation of natural urges by the moral and legal sanction of marriage in American kinship is one example, but Wagner's all-embracing account ranges over a vast cultural terrain, from ideas about artistic refinement and scientific classification to the artifices of advertising and the cultivation of personality (e.g. 1977a; 1981 *passim*).

To the extent that such varied domains of our culture (as we think of them, again, conventionally) are deemed to be different ways of rendering the world meaningful, Wagner maintains, they are also all underpinned by a particular way of thinking about meaning itself. Typified by the distinction between symbols and the things for which they stand, this 'semiotic', as Wagner calls it (1981: 42), is a corollary of the opposition between culture and nature. Meaning, according to this view, arises from human beings' ability to represent the world by bringing to bear upon it sets of arbitrarily defined (and hence conventional) symbols. Thus, it is commonly assumed, cultural conventions order the world by deploying symbolic structures that organize it into distinct categories by means of their otherwise arbitrary relationships to 'signs' – a take on meaning that is familiar to anthropologists from structuralist theory, and which endures in different ways, for example, in the otherwise divergent constructivist and cognitive approaches of contemporary anthropology (see also Holbraad 2007: 196). Indeed, it is just this view of meaning that is expressed in the basic idea underwriting the classic tradition of American cultural anthropology that the job of the anthropologist must be to 'represent' the culture he studies. Social structures, systems of exchange, modes of production, collective representations, indigenous beliefs, cultural logics, core symbols, local knowledge, cognitive schemata, logics of practice, world systems, transnational flows, political economies, multiple modernities, invented traditions – Wagner suggests that all these are ways of expressing the results of anthropologists' efforts to depict and understand the conventions of the people they study by deploying and elaborating upon their own.

This, then, is the first sense in which culture is an 'invention', as the title of Wagner's most famous and influential book has it: supposing that the only way for people to have culture is to order the world by means of conventions, the anthropologist transfigures his experience of other people's lives by using his own conventions to enunciate a set for them, too. Wagner conveys the point vividly with reference to ethnographers' experience of 'culture shock' in the first weeks of doing fieldwork, which produces in them the sensation that what confronts them is a new 'culture' to which they must 'adjust':

Anthropology teaches us to objectify the thing we are adjusting to as 'culture', much as the psychoanalyst or shaman exorcizes the patient's anxieties by objectifying their source. Once the [fieldwork] situation has been objectified as 'culture' it is possible to say that the fieldworker is 'learning that culture', the way one might learn a card game. On the other hand, since objectification takes place simultaneously with the learning, it could be said that the fieldworker is 'inventing' the culture. (1981: 8)

So, as Wagner puts it pithily, 'anthropology is the study of man "as if" there were culture. It is brought into being by the invention of culture, both in the general sense, as a concept, and in the specific sense, through the invention of particular cultures' (ibid.: 10). But of course what the anthropologist finds when living with the people he studies during fieldwork are not structures, beliefs, symbols and so on, but just people living their lives and, when he is lucky, talking to him about them. And one possibility that emerges when one attends to how people actually live is that their activities might not, after all, be directed towards establishing, abiding by, or elaborating conventions. This is precisely what happened to Wagner during his fieldwork among the Daribi (ibid.: 4–10). Accordingly, the counterpart to Wagner's argument about the sway that the notion of culture as convention holds within anthropology, and the particular assumptions about culture that it embodies, involves using his ethnography of the Daribi in Melanesia to show how culture could be conceived differently.

In *Habu*, named after a key Daribi curing ritual in which men impersonate ghosts, Wagner argues that the aspects of life the Daribi consider most salient (ritual, myth, exchange, magic, naming and more) are directed not towards controlling the world by subjecting it to collective conventions, but rather towards the opposite, namely treating their social conventions as an already 'given' baseline from which to engage in acts that are meant to transform them by way of improvisation into something novel and unique. From the Daribi point of view, all the things that the anthropologist imagines as 'culture' – 'grammar, kin relationships, social order, norms, rules, etc.' (1981: 87) – are not conventions for which people are responsible. Quite to the contrary, they are the taken-for-granted constituents of the universe that form the backdrop of human activity. They are 'innate', in Wagner's terms, inasmuch as they belong to the order of what just is rather than that of what humans have to do. Conversely, the things that the anthropologist imagines as 'nature', including not only the unpredictable facts and forces of the world around us but also our own incidental uniqueness as individual persons, for the Daribi constitute the legitimate sphere of human artifice (see also Strathern 1980). Human beings, according to this image, do not stand apart from the world, bringing it under control with their conventions, but rather partake in the world's inherent capacity to transform itself by transgressing the conventional categories that the Daribi take for granted.

So, for example, when in the habu ritual Daribi men impersonate ghosts that are held responsible for certain illnesses, they are not acting out a cultural convention – conforming to a cultural script, underpinned by indigenous categories ('ghost'), beliefs ('illnesses are caused by ghosts') and so on. Rather, like a jazz musician may 'bend' a conventional scale to improvise a solo that sounds alive and unique, they subvert 'innate' distinctions, and in this case particularly the distinction between living humans and dead ghosts, to bring about an effect that is powerful precisely because it *recasts* or, in Wagner's term, 'differentiates' the categories they take for granted (1981: 81; 1972: 130–43).

Taking as the granted state of the world that dead ghosts are dead ghosts and living people are living people (the 'collectivizing' categories of convention), in the habu men *take on* the characteristics of ghosts, temporarily enacting the startling possibility that dead ghosts can indeed come to life and interact with humans. In doing so, they artificially bring about a novel effect, namely ghosts that are also men, by temporarily transgressing ordinary distinctions between life and death, men and spirits, and so on. So, much as with jazz or good theatre acting (e.g. see Mamet 1997), the success of the habu relates to people's capacity to render the predictable unpredictable, rather than the other way round (see also Holbraad 2010). However many times the habu may have been done in the past, its power depends on the degree to which the participants can make it a fresh subversion of convention. In this sense – and contrary to anthropological arguments about ritual as a transfiguration of 'structure', 'culture' or 'ideology' (e.g., Geertz 1973; Sahlins 1985; Bloch 1992; Rappaport 1999) – the habu is an anticonvention par excellence, or, in Wagner's word, an *invention* (see also Wagner 1984; Strathern 1990).

So, for Wagner, the second sense in which culture is invented presents itself in direct contrast to the first. Anthropologists invent a culture for the people they study in assuming that what makes them different (viz. an example of 'cultural variation') must be the particular way in which they organize their lives conventionally. By contrast, the Daribi are different in that their energies are focused on 'differentiating' their conventions so as to bring about singular moments of invention. Hence if our slot for 'culture' is the slot of what people 'do', and our slot for 'nature' is for that to which they do it (see Figure 2.1), then in the case of the Daribi the slot for 'culture' is taken by the activity of invention and that

	for anthropologist	*for Daribi*
innate ('nature')	invention	convention
artificial ('culture')	convention	invention

Figure 2.1 Wagner's nature/culture reversal.

of 'nature' is taken by 'innate' conventions. In that sense Daribi culture *is* invention.[4]

The question now arises: if what counts as culture in the case of the Daribi is in this way opposed to what counts as culture in anthropology, then what could an anthropological account of the Daribi look like (or for that matter a Daribi account of anthropology – Wagner 1981: 31–4)? In other words, if what makes the Daribi 'different' is their orientation towards invention, then how can an anthropology that brands all differences as divergences of conventions ('cultural variation') make sense of Daribi life without inventing for it a 'culture' that distorts it? This goes to the crux of the central question that we posed earlier: how can (in this case) the Daribi's radical divergence from our distinction between nature and culture be articulated without falling into the Schneiderian trap of thinking of it as 'cultural'?

---

[4] The sense in which culture is an invention for Wagner is directly opposite to the sense in which it can be, famously, for Hobsbawm and Ranger, when they write of 'invented traditions' (1983). Hobsbawm and Ranger's core observation is that practices that may seem 'traditional' typically turn out to be recent inventions that act to legitimate present practices by establishing a sense of continuity with an often-fictitious past. In terms of Wagner's argument, the claim amounts to the idea that conventions typically purport to be older than they actually are. What is being invented for Hobsbawm and Ranger are not inventions in Wagner's sense, but rather new and suitably old-looking conventions. While the 'invented traditions' argument is persuasive in its own terms, (though see also Sahlins 1999), it should in no way be confused with Wagner's, which is that the very assumption that people are bound always to control the vagaries of history by appeal to the stability of convention (putative or otherwise) may in some cases have more to do with the analyst's needs than with those of the people he studies.

Wagner's solution to the conundrum relates to the third and final sense in which culture can be conceived as an invention for him. This is, effectively, an exhortation to move from the first sense in which culture has always been an invention for anthropologists (the idea that making sense of other people's lives must come down to formulating – and in that sense inventing – the conventions by which they live) to the second one (the Daribi-derived idea that human activities may be oriented towards subverting conventions in order to precipitate singularly novel effects). As Wagner writes:

Anthropology will not come to terms with its mediative basis and its preferred aims until our invention of other cultures can reproduce, at least in principle, the way that those cultures invent themselves. We must be able to experience our subject matter directly, as alternative meaning, rather than indirectly, through literalization or reduction to the terms of our ideologies. (1981: 30–1)

Thus, if the assumption that people like the Daribi must have a culture that consists of conventions gets in the way of making sense of the fact that culture in their case consists of processes of invention, then the onus is on the anthropologist to move away from his initial assumptions and conceive new ones, to avoid reducing them 'in terms of our ideologies'. Hence the ethnography of Daribi 'invention' must precipitate a process of invention *on the part of the anthropologist*. Departing from the conventional anthropological notion of culture as convention, the anthropologist is called upon to transform the notion in a way that incorporates the possibility of invention as described for the Daribi – in other words, to re-invent the notion of 'culture' as invention.

Set forth as the 'epistemology' of the ethnographic argument presented in *Habu* (Wagner 1981: xv), *The Invention of Culture* does just that. Putting in place the conditions of possibility for invention on the part of the anthropologist, much of the argument of the book is devoted to showing that such processes of invention are as present in modern Western life as they are in Melanesia (a territorializing distinction between 'the West' and 'Melanesia' – or 'tribal people' more generally – that Wagner uses rather broadly throughout his work, as we shall be discussing critically towards the end of the chapter). The difference as Wagner sees it is that whereas 'Melanesians' take responsibility for their acts of invention while taking conventions for granted, for 'Westerners' it is the other way round. Jazz and acting, mentioned earlier, are only particularly 'marked' examples of this – displaying not only the presence of invention in our lives in the West, but also its relegation to an irreducibly ineffable realm that lies beyond human control (notions such as 'natural talent', 'inspiration' or even 'genius', meaning 'something ... je ne sais quoi', bear the point out). In fact, argues Wagner, the possibility of invention is implicated in any action or statement whatsoever (1972: 8–9; 1981: 34–55). Even the simplest declarative sentence, insofar as it is not just a trivial statement of what is already known conventionally (e.g. literal statements about accepted categories, such as 'men are mortal', or definitions such as 'bachelors are unmarried men'), involves an element of invention. For example, when we ourselves, here, say that invention is part and parcel of all non-trivial communication, or if we suggest that Wagner's analysis is elegant, what we say is only interesting – it 'says something' – insofar as it presents a *subversion* of what is already conventionally accepted. We are in effect asking you, the reader, to conceive of communication and of Wagner's analysis differently from what (we assume) you already do. In this sense we are putting forth an invention in a way that is analogous to what Daribi men do when they impersonate ghosts (and compare this with such exemplarily uninteresting claims as 'communication involves mutual understanding' or 'Mozart's music is elegant' – boring, precisely, because they say nothing 'new': they merely state conventions that are already established and accepted). The same, argues Wagner, holds for all meaningful action that is not merely trivial (1972: 4–5; 1981: 39–41).

Just as with conventional accounts of rituals such as the habu, the problem, Wagner suggests, is that the irreducibly inventive dimension of meaning remains opaque as long as it is viewed through the prism of convention. To equate the meaningful with the inventive is itself an example of invention insofar as the assumption 'against' which this point gains its originality is that meaning pertains to convention – the 'semiotic' of symbols that stand for things in the world that Wagner identifies, as we saw, as the key corollary of the standard anthropological distinction between culture and nature. So if it is to be understood as more than such an act of trivial classification, the notion of invention requires an alternative account of meaning. Such an account lies at the heart of Wagner's own invention of the notion of invention and takes the form of a dialectical contrast with the semiotic of convention – a 'figure/ground reversal', as he often puts it (e.g., Wagner 1987). Let us explain precisely how.

Convention relies on the assumption that the realm of symbols and the realm of the things for which they stand are opposed – culture to nature, representation to world. Conventions arbitrarily 'fix' the meaning of symbols that can then be used to express things by being 'applied' to the world. For example, we assume the meaning of 'elegant' and 'Wagner's analysis' is already clear, so when we say 'Wagner's analysis is elegant' we apply the former to the latter – an operation that is formally indistinguishable from our doing the same for Mozart's music. Hence, in the semiotics of convention, the fixing of meaning and its application to the world to express something 'about it' are logically separate, the former being the precondition for the latter. Invention, Wagner argues, turns this image inside out. When the Daribi impersonate ghosts, a jazz musician goes off on a solo, and Wagner invents the semiotics of invention (and we call it elegant), meaning is not a precondition for expression but rather an *outcome of it*. According to this view, every act that the semiotic of convention would brand as an 'application' of symbolic meaning to the world is in fact an *extension of meaning*.

So, to stick to the example, to say 'Wagner's analysis is elegant' does not just 'apply' the notion of elegance to Wagner's analysis but also, in doing so, extends what we mean by both: the particular qualities of Wagner's analysis putatively extend the notion of elegance in a novel way and, equally, the notion of elegance extends the way we think of

Wagner's analysis. Instead of a 'gap' of mutual independence between symbol and thing, representation and world, we have a relation of mutual dependence, whereby meanings modify each other in the act of being brought together. Just as with the habu, then, the semiotic of invention implies that everything ('representations' and 'world' alike) is meaningful, and the task of expression is to mediate the relationships between meanings so as to engender novel ones – not to 'convey' meanings representationally, but to *create* them by transforming the ones that are already given.

We may conclude, then, that if making sense of the Daribi's divergence from our anthropological assumptions about nature and culture precipitates an act of invention on the part of the anthropologist, it must also precipitate *a departure from just those assumptions*. The anthropologist's task of making sense of people that are 'different' can no longer be a matter of 'representing' the differences in question – to do so, as we have seen, is effectively to obliterate them. Rather like the habu, the anthropologist's task on this view is to transform the categories he takes for granted in the very act of bringing them to bear on the differences of which he seeks to make sense. Making sense, in other words, must involve the semiotics of invention.

This way of thinking of the task of anthropology 'solves' the Schneiderian paradox. Instead of trumping ethnographic alternatives to the distinction between nature and culture by branding them as cultural, Wagner's approach allows them to trump that distinction itself by transforming it into an altogether different way of thinking about difference. To think about difference, on this view, *is to think differently*: to transform one's most basic assumptions in light of the differences that trump them. We have here, then, nothing short of an anthropological (indeed, ethnographically driven) derivation of the central tenet of the ontological turn as we understand it – the central idea that has been taken up and explored more and more explicitly in the decades that followed Wagner's conceptual breakthrough.

The Obviation of Meaning

To deepen our understanding of Wagner's theory of invention and its decisive impact on what later developed into the ontological turn in anthropology, we now turn to his studies of myth, which he considers as the cultural form in which the workings of invention render themselves most explicit. We focus particularly on *Lethal Speech* (1978), the sequel to his theory-of-everything *The Invention of Culture*, in which Wagner returns to his Daribi ethnography in an effort to develop and further specify the implications of his general theory of invention with reference to the study of Melanesian myths. As we shall see, the upshot is a model of mythical invention that unpacks in great detail the manners in which the process of invention is able to develop and further specify its own implications, through what Wagner brands technically 'obviation sequences'.

For Wagner myth is distinguished by the complete manner in which it 'spreads out' (1986a: 34) the interplay between invention and convention. To see how this process of spreading out unfolds, therefore, we may start by returning to Wagner's basic contrast between these two semiotic modalities, and particularly to the manner in which they come to relate to each other alternately as figure to ground. Consider once again the startling proposition of the habu-dancers, 'we are ghosts'. As we saw, this is inventive insofar as it takes conventional assumptions for granted (e.g. 'we are human') and, working against them, transforms the meaning of the terms involved. Within the simple declarative statement 'we are ghosts', then, is to be found a form of movement, namely a shift of meaning. Spelling this out involves isolating three mutually implicated steps.

First, as we saw earlier, we have what we may call the default position, according to which humans and ghosts are assumed to be distinct kinds of being – humans are humans and ghosts are ghosts. This is the conventional starting-point, which fixes the coordinates against which the habu-statement 'we are ghosts' gains its meaning. Then we have the second step, in which the conventional assumptions are disrupted in order to be refigured in a unique, 'differentiating' way. This is the movement of metaphor in the etymological sense: the notion of being a ghost is 'carried across' (*metafora*) onto the notion of what 'we' might be, transmuting itself across the conventional divide between ghosts and humans in what Wagner elsewhere calls the 'analogic flow' of meaning (1986a: 18; 1977b). So, as Wagner puts it technically, while the default distinction between humans and ghosts serves to 'facilitate' the invention by providing the initial coordinates that the habu-statement disrupts, this second step serves to 'motivate' the invention, not only by providing a 'reason' for the disruption (viz. an alternative possibility of meaning), but also because it enacts it as a movement, namely the flow through which the meaning of 'humans' and 'ghosts' enter in a relationship of mutual reconstitution. The effect of this motion, then, constitutes the third step of the invention, in which this movement of meaning is arrested so as to yield a new set of semiotic coordinates. The possibility that, while human, the habu-dancers may also become ghosts is crystalized as a shift of the initial coordinates of meaning, so that the conventional distinction between humans and ghosts is supplanted by the habu-proposition 'we, as humans, are ghosts' (a proposition that differentiates the habu-dancers from the rest of the participants in the ritual, and not least the ailing victims for whom the ritual is held, who remain vulnerably human).

Adapting the many diagrams of triangles that populate the pages of Wagner's exposition of the method of obviation in *Lethal Speech* (1978) as well as his later development of the model in *Symbols that Stand for Themselves* (1986), Figure 2.2 presents the three steps of metaphoric invention in the form of a triangular sequence, ABC. Since for Wagner obviation sequences are essentially manners of 'spreading out' the basic pattern of metaphor, to understand how obviation works it pays to start by thinking through some of the implications of its triangular structure.

Wagner's own explanation of why metaphors are best thought of as triangles is couched in the language of dialectics. More than just an

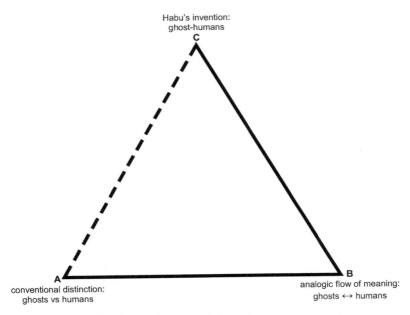

Figure 2.2 Wagner's scheme of metaphoric invention.

opposition, the relationship between convention and invention that metaphor embodies is one of mutual mediation that closely parallels a classic Hegelian trio of thesis, antithesis and synthesis (although we shall see that in a crucial way it is different). The default distinction between humans and ghosts (call it thesis A) is contradicted by the 'differentiating' flow of analogy that runs the two together (antithesis B), and then the contradiction is resolved by the new settlement of meaning that the habu-proposition expresses (synthesis C). The dialectical (as opposed to just binary) nature of the relationship between convention and invention is owed to the mutual mediation between them. If the metaphoric flow that allows the meaning of 'human' to be modified by that of 'ghost' is the heart of the habu-invention (antithesis B), that is because it mediates (and motivates, as we saw) the shift between the default assumption (thesis A) about the meaning of these terms and their novel redefinition in the habu (synthesis C). Conversely, looking at this process from the viewpoint of its outcome, the habu-statement C mediates between thesis A and its antithesis B in that it is the result of the modification of the former by the latter and is in that sense their 'synthesis'. So far, so Hegelian.

However, to these two forms of mediation (call them syntagmatic and paradigmatic respectively) Wagner adds a third, which, while not necessarily contradicting the Hegelian model, takes the notion of dialectic progression in a startlingly novel direction, namely *backwards*. For if one is prepared to think against the flow of invention (from C to B to A), the initial point of departure of the invention, thesis A, can also be understood as having a mediating role. As Wagner puts it, it provides the 'context' (cf. 1981: 37–50) with reference to which the particular settlement of the flow of meaning that C embodies makes sense. It is only by referring *back* to the initial distinction between humans and ghosts (A) that the habu-statement (C) 'we are ghosts', which is made possible by the analogic flow of meaning (B), makes sense at all. Wagner's technical term for this form of retrospective mediation is 'counter-invention' (ibid.: 47), drawing attention to the fact that every act of invention ('we are ghosts') involves *ipso facto* also a retroactive re-invention of the very meanings upon which it innovates ('humans', 'ghosts'), figuring them as the 'obvious' grounds that need to be overcome as and when new meanings are, in turn, refigured upon them. In the very act of establishing that they are ghosts, the habu-dancers must willy-nilly remind us, as it were, that normally we'd take them to be human *instead*. In this way the habu 'synthesis' C both stems from the default thesis A and takes us back to it, thus closing off the triangle, if only by implication and largely as a side effect (hence the intermittent line connecting C to A in Figure 2.2).

This, then, is the point of *obviation*. As the three-step pattern of invention illustrates in rudimentary form, acts of symbolic/metaphorical expression – and for Wagner *all* acts of expression are symbolic and metaphorical to some extent – involve dispensing with prior meanings, taken for these purposes as already established, in order to supplant them with new ones. So in both senses of the word, and with a strong tinge of paradox, meaningful expression for Wagner is an act of obviation: it requires that already established meanings are revealed ('made obvious') as being unnecessary, overcome, old hat, 'obviated'. Speech (i.e. meaning) is indeed *lethal*, as the title of Wagner's book would have it, and perhaps the best image of its motion would be that of a snake eating its own tail (although to do justice to Wagner's conception we would have to imagine the snake forming a triangle rather than a circle). And it is on this point that Wagner's model of dialectic differs quite markedly from the Hegelian image. The latter posits dialectical motion as an extensive progression, repeating itself linearly as each successive synthesis becomes the point of departure for a further dialectical contradiction, requiring a further synthesis, if not ad infinitum, then perhaps till the end of history. Wagner's model of obviation, by contrast, posits the motion of meaning as an irreducibly intensive and 'non-linear' process – a matter of 'unpacking' the density of the flow of meaning in determinate ways, rather than extending it by adding to it new elements from beyond itself. Meaning mined – 'elicited' – out of its own resources. Symbols, as the title of his famous book has it, that stand for themselves (1986).

On Wagner's model, then, the backwards-and-forwards (vicious) circularity of traditional hermeneutics – wholes presupposing parts and parts presupposing wholes – is replaced by the forward-thrusting motion of a spiral: meaning that closes in on itself ever and again only by moving forward (see Box 2.1). Here the generation of meaning depends not on aggregating and disaggregating parts to make or break up wholes, but rather on the possibility of what we may call 'auto-substitution', with semiotic wholes generating ever new wholes out of themselves, either by unfolding themselves outwards (invention) or by folding inwards, back onto themselves (counter-invention, viz. obviation). Indeed, as James Weiner points out in one of his many illuminating commentaries on, and extensions of, Wagner's analysis, 'the folding of imagery back on itself so that it comes to have an inside and an outside' (1995: xviii; see also 1988; Wagner 2000) is as good a way as any to define Wagner's difficult but crucial idea of obviation.

Box 2.1: Part-whole relations in obviation and the hermeneutic circle

The standard idea one associates with the hermeneutic circle is that, when it comes to interpreting the meaning of any particular expression, the whole and its parts are related internally, i.e. the whole can only be understood in terms of its parts and vice versa. We may run the thought on our example from the habu: the meaning of the statement 'we are ghosts' can only be understood with reference to the meaning of its constituent parts, 'we' (e.g. as humans), 'are', and 'ghosts', and, conversely, each of these words must be understood (at least partly) with reference to the meaning of the larger statement, 'we are ghosts'. Parts and wholes, as it is often put, provide a 'context' for each other. As is also often said, however, if this image presents interpretation as a circular process, tacking back and forth from whole to parts in order to make sense of either, then it also presents it as a problem, since the circle seems to be vicious: if to understand the habu-statement I must first know what the words it brings together mean, and to know the meaning of those words I must already know what the habu-statement means, then how can I ever know either? An old conundrum, not least for anthropologists (Geertz 1983: 69).

Wagner's model of meaning is free of this problem because it configures the relationship between parts and wholes quite differently. To say, as in the hermeneutic formulation, that to understand the whole one must first understand the parts and vice versa is to imagine the relationship between the two as one of aggregation or disaggregation of units, a bit like one might imagine the relationship between a completed jigsaw puzzle and its constituent pieces: to

---

complete the puzzle is a matter of adding together in the right way pieces that are already available as ready-made units while, conversely, to see where each piece might fit is partly a matter of notionally breaking down to its constituent parts the image already given on the cover of the box. By contrast, Wagner's model of meaning does not depend on adding things up or breaking them down into their pieces. In place of aggregation and disaggregation he posits an economy of generation and destruction. In the habu, as we saw, understanding the meaning of 'we are ghosts' is demonstrably not a matter of stringing together meanings that are already available ('human', 'ghost' etc.), but rather one of substituting – destroying, exhausting, 'killing' – those meanings by transforming them dialectically into new ones. Far from presupposing the meanings of 'human', 'ghost' and so on (the position of 'convention' in Wagner's terminology), the habu-statement 'we are ghosts' invents them and, in the process, also 'counter-invents' the original assumptions against which this invention makes sense, retrospectively bringing about the very 'parts' of meaning that hermeneutics takes for granted as 'units'. So the relationship between parts and wholes becomes one of mutual generation, rather than the (viciously circular) mutual presupposition of the hermeneutic model. It is as if each piece of the puzzle already contained within itself the whole image on the box and, conversely, that whole image projected its putative pieces outwards into space, each piece standing alone as a whole image in its own right.

In subsequent works Wagner himself has drawn out a series of startling consequences of this way of thinking, including the mind-bogglingly 'holographic' effects, as he calls them (2001, 2010), of what is effectively a three-dimensional model of meaning. The language of non-linear systems (2001), chaos theory (2005) and fractality (1991) has been

Analogic Anthropology

a central feature of these writings, and this is also one of his main points
of contact with the work of Marilyn Strathern, as we shall see in the next
chapter. Indeed, insofar as the ontological turn in anthropology is characterized by its propensity for conceptual innovation and neologisms, a
telltale sign that Wagner can rightfully be considered its progenitor is the
sheer mass of concepts he has developed over the years. As with obviation, many of them are about adding a dimension of depth to the concept of meaning: figure/ground reversal (1987), the outward symmetry
of twinning (2001: 48–63), expersonation as opposed to impersonation
(2010: 47–102; 2012b), and more. When Wagner talks of 'holography' he
means it: his game in anthropology, one might say, is to create conceptual holograms. For present purposes, however, we restrict our account
to Wagner's original deployment of obviation in the analysis of myth,
since it is in the guise of what Wagner calls myth's 'obviation sequences'
that the core tenets, manners of operation and sheer analytic power of
his model become most explicit.

Myth and Its Obviation Sequences

Myth is the quintessential obviational form. For while the dialectic of
convention and invention upon which obviation turns is at play in every
meaningful aspect of life – indeed, we are performing continual acts of
obviation right now, as we write this, and so are you, as you read what
we wrote – in myth this process becomes fully transparent since, more
than any other form of expression (barring, possibly, jokes – see Wagner
2001, 2010; Pedersen 2011: 183–205), for Wagner myth is 'about' its own
meaning (1978: 13). Echoing Lévi-Strauss's famous contention that myths
are not ultimately about anything other than 'the mind that evolves
them by making use of the world of which it is itself a part' (Lévi-Strauss
1969: 340, cited in Wagner 1978: 52), Wagner conceives of myth as a 'self-
contained and self-generative' realm of meaning that is 'as much "about"
itself as it is relevant to the study of a culture' (1978: 13). Its 'facility', he

writes, 'is not that of replicating the world, but of setting up its own world in contradistinction' (ibid.: 33).

The crucial difference between Wagner and Lévi-Strauss, however, is that for Wagner the self-contained and self-generative character of myth is not owed just to the freedom with which myth deploys, mediates and thus proliferates structural oppositions, as Lévi-Strauss's account of mythical meaning would have it (e.g. Lévi-Strauss 1969). For Wagner, to view myth as this kind of free combinatorial operation – as a form of 'mental gymnastics', as Lévi-Strauss famously put it (1969: 11; cf. Friedman 2001) – is effectively to contain it within a semiotic of convention, as if all myth did were to play around with the structural possibilities allowed for by an already existing – 'underlying' – code of oppositions (nature/ culture, raw/cooked, hot/cold etc.):[5]

[I]t must be recognized that [a myth's] dramaturgical order is one of successive *changes* or *displacements*, in which social or cosmological elements acquire significances intrinsic to the story itself. The charting or ordering of such elements, as Lévi-Strauss and others have done, is perhaps a kind of discovery procedure, but it is deceptive in that the elements are significant in the myth precisely as they diverge from conventional uses. (Wagner 1978: 13)

Thus adopting Lévi-Strauss's model is to ignore the other half of the story, namely how myth is able to constitute itself as 'its own world in contradistinction' to conventional orders of meaning *by inventing itself against them*. If 'every myth is a unique experiential world', Wagner writes, that is because 'substitution invariably changes, extends, and relocates the recognized differences and similarities, displacing through its own creative action any possible lexical guides' (1978: 38). What myth does, in other words, is to take conventional orders of meaning and

---

5   While the contrast with Lévi-Strauss is a recurring theme in Wagner's work on myth (e.g. 1978: 35–7, 51–2; 1981: 150–1; 1986a: 131), perhaps the most detailed and systematic comparison between obviation and structural analysis is offered by Weiner (1988: 154–72).

substitute them dialectically for invented ones, revealing, in the process, 'the range of possibilities opened up by symbolic innovation' and, by the same token, providing a 'glimpse of the limits of social protocol' (Weiner 1995: 37–8). Myth, in that sense, is one big act of obviation.

Myth's self-contained, world-within-world character lies in the manner in which it makes its own process of obviation explicit. Exemplifying the holographic logic of auto-substitution, it does this by 'spreading out' the unfolding and enfolding movements of obviation as a series of steps in the myth's 'plot'. An initial situation is set up as the myth's conventional point of departure, in the manner of 'once upon a time…' Then a series of incidents in the plot serve to erode this opening situation, gradually displacing it, distorting it or rendering it otherwise problematic (Weiner 1995: 38). Wagner conceptualizes this dramaturgical progression as a series of 'substitutions', i.e. with each new incident in the plot displacing the previous one (e.g. first the heroine is at home with her parents, then her mother dies and the cruel step-mother moves in). Exemplifying the logic of obviation, these substitutions alternate dialectically between the 'facilitating' conventional circumstances established by the opening situation and a series of events that disrupt it, inventing against it incidents that 'motivate' reciprocal readjustments of the conventional order that move the story forward (e.g. the arrival of the step-mother forces the daughter to take the role of a maid). Each readjustment displaces the opening premise of the myth a little further, culminating in a final substitution that explicitly contradicts the initial convention, effectively making it collapse under the weight of the myth's plot (e.g. a slipper left at the royal ball is found to fit the right foot, so the heroine, no longer wretched, marries the prince and becomes a princess). This is the moment of obviation, where the dialectical sequence of the myth's plot closes in on itself, eating its own tail, having completed its whole-for-whole auto-substitution.

In line with Wagner's holographic model of meaning, such obviation sequences effectively render myths three-dimensional, adding depth

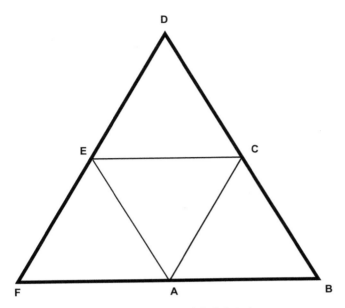

Figure 2.3 Wagner's holographic model of obviation.

to them by showing the 'inner workings' of the auto-substitutions that they perform. Conveying their 'fractal' structure, Wagner models these sequences by taking the basic triangle of obviation and multiplying it by itself holographically, which is to say inwards, as shown in Figure 2.3. A single overall obviation (A to F), equivalent in its closure to our earlier example of the habu-statement, is revealed as containing within itself three interlocking sequences of auto-substitution (ABC, CDE, EFA), each of them reproducing 'fractally' or 'holographically' the triangular form of the obviation. The symmetry produced by this structure reflects the dialectical alternation between the 'facilitating' series of conventional circumstances that are set into motion by the 'motivating' series of events that invent against them – a binary interplay that takes the ternary form of obviation, with two triangles mediating each other (ACE facilitating, BDF motivating).

Throughout his writings, starting with *Lethal Speech*, Wagner has used this basic triangular structure to dizzying effect to model different levels of complexity within and across myths (e.g. 1978: 49–51, 144–9), as well as their links with ritual (e.g. 1986a: 58–80; 1986b: 211–20) and even history (e.g. 1986a: 96–125; see also Nielsen 2011; Pedersen & Nielsen 2013; Holbraad in press a). One might object, of course, that myths themselves rarely offer themselves up in the form of such elegant symmetries. Indeed, as Lévi-Strauss also acknowledged, myths are full of redundancy and apparently wilful disjointedness (1963: 229). Steering clear of such charges from misplaced concreteness, Wagner makes clear that the triangle of obviation sequencing is an analytical device rather than a descriptive procedure:

The diagram is not intended to represent self-evident or empirically discoverable motifs or segments of a text; it is not presented as an inductive summary of the plot, a *hypothesis* regarding the mythmaker's intentions. It is, rather, a deductive construction depicting the implications and interrelationships of a set of events that are themselves relations (substitutions). The substitutions that I identify with this schema may be more or less obviously featured in the text, yet the important thing is not the (literal) closeness of 'fit', but the rapport that the interpretation finally achieves with the sense of the tale – the degree to which the interrelationships make 'sense'. What is diagrammed, then, is not the shape of a myth, but the shape of my interpretation. (Wagner 1978: 48; emphasis original)

Recalling our discussion in the Introduction of the ontological turn's deliberate propensity to 'fail' with respect to standard protocols of hermeneutic exegesis in anthropology, Wagner's commentary on his own mythical interpretations prefigures the characteristically experimental way in which the ontological turn seeks to refigure the relationship between ethnographic materials and analytical procedures. To give a sense of what such interpretations look like in Wagner's work on myth, we now briefly consider his own original illustration of obviation in the opening chapter of *Lethal Speech*, with reference to a Daribi myth[6] that relates the origin of food crops to an act of human parricide. We paraphrase the relevant part of the myth (cf. 1978: 39–51):[7]

Once an old man was living together with his two sons. Lacking edible food, they had to live on tree fruit [which is what gardening birds are deemed to garden and eat]. The eldest son went in search for food. Encountering a gardening bird, he tries to shoot it but misses, his arrow making contact only with its feathers. The feathers turn into food-domesticates of various kinds (sweet potatoes, sago, bananas etc.), which the man collects and takes home. He asks his younger brother to cook them and then goes out to find the bird again. Losing track of it, he finds himself at the clearing where he had earlier shot at it, only now it's a garden. There the bird appears in human form and recounts to him how he too (the bird) had lived here with his own father without food, and that when his father had died food-crops had sprouted from his burial place. So he gives the man foods to take with him, telling him that when he gets home he should ask of his brother that they kill their father and bury him. When the man got home he said this to the brother and the two of them were despondent and hesitant. Still, in his sorrow, the eldest brother kills the father, buries him, and out of the burial place sweet potatoes, sago, banana and other foods begin to grow.

Figure 2.4 maps the turns of the myth's plot onto Wagner's obviational model – his triangle of triangles. Without going into the detail of each

---

[6] We could have equally well used our standing example of the habu-statement, since Wagner has provided a detailed obviational analysis of the whole sequence of the habu curing ritual itself (1986: 69–80), thus illustrating how obviational analysis can be extended beyond the plot-lines of myth. We have chosen to illustrate obviation sequences with reference to myth, however, staying close to Wagner's own original conception.

[7] For purposes of this brief illustration, in our paraphrase we include only those parts of Wagner's transcript of the myth that feature in the obviation sequence he extracts from it interpretatively. Curious readers are encouraged to consult the original transcript (1978: 39–45), where they will find, for example, that the myth continues with some further episodes, which Wagner interprets as an *apologia* that, as he proposes, extends the obviational sequence by adding to it its own inversion, revealing what Wagner calls an 'internal myth' (ibid.: 48–51) – this being an example of the kind of complexity Wagner builds into his obviational analyses.

step in the sequence (see ibid.: 45–8 for Wagner's full analysis), the over-all picture here is of a movement from an opening situation (A) to a final outcome (F). To start with, humans (presented as a microcosm of father-with-sons) lack edible food, having to make do with tree-fruit like the birds. By the end, humans are able to enjoy gardened food, but in order to do so must first (E) engage in a wretched act of parricide, thus institut-ing the social form of paternal succession. This overall movement is one of obviation: a situation set up as the myth's conventional point of depar-ture, namely a time in which humans had to make do with wild food crops, is ultimately overcome by the advent of a world in which food cultivation is a human activity. This technological transition, accord-ing to the myth, is a function of a social one: from a time before time, when fathers and their sons lived alongside each other (presumably) in perpetuity, to a time when, like now, fathers die and are succeeded by their sons.

So, Wagner formalizes the obviation produced by this complex transi-tion as a substitution of relationships: in the initial convention fathers and sons stand together and in contrast to the undomesticated food crops (father + sons / tree food), while with the invention of domesti-cated food the sons stand together with the food crops and in contrast to the father whom they killed in order to procure them (sons + garden / father). Much as with the habu-statement (there the substitution would be humans / ghosts → humans + ghosts), we have here a startling shift in the coordinates of meaning, where 'fathers', 'sons' and 'food' and the relationships between them are all drastically redefined. Note, however, the difference from habu: While there the goal of the obviation was to take a social convention and invent a powerful new effect out of it – men that are ghosts –, here we have effectively the 'origin story' of what is an established social convention 'in real life'. The goal of obviation, then, is to give us the back-story that *leads up* to social convention, making it apparent that what 'we', the Daribi, take for granted is actually an inven-tion that emerged out of particular circumstances 'once upon a time'.

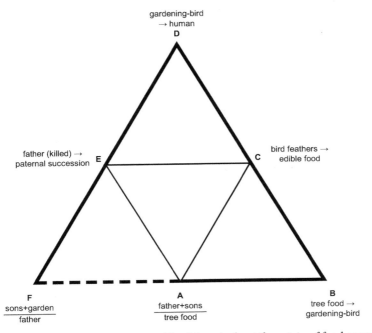

Figure 2.4 Obviation sequence of Daribi myth about the origin of food crops.

Hence the originary situation is set up as an older convention against which what is now taken to be conventional – domesticated crops and paternal succession – was actually invented, this realization lending the story its power.

The redefinition of meaning with which the myth culminates, then, is achieved piecemeal, step by step, as the opening convention is gradually eroded by the episodes that succeed it. In particular, the dialectical interplay between convention and invention through which this shift of meaning proceeds takes the form of, on the one hand, 'facilitating' circumstances to do with the relationship between the father, the two sons and their search for food, and, on the other, a series of 'motivating' encounters with gardening-birds and the horticultural knowledge that they are able to bestow: humans' socio-culinary circumstances (convention) are gradually modified, adjusting to the events involved in humans' encounters with the birds (invention). With each dialectical adjustment the story takes us further away from the initial circumstances and closer to the final event that obviates them. This is achieved through a series of further substitutions that gradually transform the initial premise of the story in the direction of its resolution.

For example, against the background of the initial convention (father + sons / tree food), Wagner formalizes the first event of the story (B), when the son shoots the gardening-bird, as a substitution of one food for another (tree food becomes gardening-bird). This event then precipitates an 'adjustment' of convention (C) when the bird's feathers become (i.e. in Wagner's analytical vocabulary are 'substituted by') edible foods. So already, with this first triangle of obviation, we have a significant shift in convention, from an original, 'pre-human' situation of people eating tree fruits like birds to an intermediary situation where, thanks to the birds, they acquire human food for the first time. This new situation then facilitates a further event of substitution (D: the bird reveals itself as a human too), followed by another conventional adjustment (E: the parricide that inaugurates the new social form of succession), which in turn sets the conditions for the final substitution (F: garden for father), with which the obviational sequence closes itself off.

By way of closing our account of Wagner's account of meaning, we may note how similar his obviation sequences are to Lévi-Strauss's structural analyses of myth: the emphasis on formalization, binary relations, their successive substitutions and so on. Indeed, in Chapter 4 we shall see that Lévi-Strauss's abiding concern with the transformation of meaning, which Wagner shares, has had a pervasive influence in the development of the ontological turn as a whole, particularly in the work of Viveiros de Castro. In line with our earlier contrast between the two great mythologists, however, we may also note an important sense in which Wagner has the edge in the analysis of myth, at least when it comes to accounting for myths' characteristically story-like form. The core structuralist notion that myth is ultimately an expression of an underlying structural matrix tends to pull Lévi-Strauss's analyses away from the overt dramaturgical dynamic of mythical narratives, treating it as an epiphenomenon of a 'deeper' set of structural relationships. Drama is merely the wrapping, structure is the sweet. By contrast, Wagner's sequential treatment makes virtue – and sense – of the drama. The successive mediations that constitute the myth's 'inner world' lend its narrative a story-like sense of direction, while the dialectical character of this 'motivation' ensures that the story's direction remains open at any given point. The plot's turns seem predictable (even 'obvious') only in retrospect, as an effect of the dramaturgical closure that the moment of obviation brings – this being the narrative's punch line or climax, where the 'point' of the story is finally revealed (indeed, Wagner often points to the analogies between the climax of mythological sequences and events and the punchline of humorous sequences and events – e.g. see 2001).

And the revelation is a profound one, not least for the people listening to the story. Characteristic of societies in which, as we saw, conventions are taken for granted as innate, the net effect of myths such as our Daribi example is to reveal that, actually, conventions too originated as inventions. Gardened food and patrilineal succession may seem part of the necessary order of things, but, as the myth shows, it turns out they came about contingently through actions and incidents past. Conversely, as Wagner shows for the Daribi elsewhere in *Lethal Speech*, the revelation can also work in the opposite direction, with artificial human inventions (e.g. a tale spun out of entertaining incidents) placed in the 'facilitating' role, and then gradually revealed as being subject to the innate constraints of convention (e.g. collective moral strictures) that act as the 'motivators' of the story – Aesopian morality tales are an example. In either case, the dialectical conversion of convention into invention and vice versa, which mythical obviation performs, has what Wagner calls a 'relativizing' effect (1981: 54–7): what we take for granted is unsettled by the revelation that it could have been otherwise, while our sense of personal responsibility and capacity for achievement is qualified by the realization that things ultimately just are as they are.

Much of Wagner's work can be seen as an attempt to 'parse' this basic insight, about the revelations the interplay of convention and invention produces, for an array of different cultural forms: ritual, visual art, music, poetry, history, gambling, advertising, jokes, science and science fiction, his ongoing fascination with the works of Carlos Castaneda, and much more. Just like different genres of myth, each of these cultural forms may configure differently Wagner's core distinctions between the innate and the artificial, facilitation and motivation, differentiation and collectivization, and convention and invention, and develop them afresh. Taken as a whole, however, his work builds up what one might call a consummate poetics of revelation: what we make is revealed to us, and revelations are of our own making. Either way, what the dynamic of Wagner's obviational dialectic reveals *analytically* is that the relativizing possibility of thinking afresh what one thought one already understood is built into the very infrastructure of meaning. It is of the essence of meaning as a phenomenon, if one may speak in these terms, that it contains within itself the horizons of its own renewal.

Wagner's Ontology

Creating an anthropology that contains within itself the horizons of its own conceptual renewal is as good a way as any to describe the ontological turn's aspiration for the discipline. So, might one imagine Wagner's systematic account of the self-renewing character of meaning as providing the theoretical foundations for the ontological turn? Certainly, Wagner's argument lends itself to such an interpretation. As we have already seen, Wagner's conceptualization of invention demonstrates in an exemplary way the central methodological injunction of the ontological turn: confronted with Daribi lifeways, the anthropologist is forced reflexively to reconceptualize one of the prime conditions of possibility for the anthropological encounter itself, namely the notion of cultural difference. Furthermore, the fact that the difference or 'alterity' in question resides precisely in the way Daribi transgress anthropological expectations of what 'culture' might be allows Wagner to use the content of his ethnography as a resource for transforming the form of its anthropological analysis: his reinvention of 'culture' as an anthropological concept proceeds by comparatively transposing his ethnographic argument about the role of invention in Daribi onto the level of anthropological analysis, demonstrating how ethnographic exegesis itself is an example of the very workings of invention that Daribi lifeways reveal.

In the Introduction we explained the sense in which such acts of anthropological invention can be understood as 'ontological', and Wagner's strategy provides an excellent example: his invention of culture in the light of Daribi manners of invention amounts to a concerted redefinition of what 'culture' is (i.e. culture as convention and culture as invention are *two different things*). Indeed, one might say that his development of the semiotic of invention provides nothing short of a systematic conceptualization of conceptualization itself, understood as an act of ontological transformation – shifting, if you like, the coordinates of the meaning of meaning.[8]

---

[8] In particular, Wagner's theorization of metaphor and obviation can be read as an account of the inner workings of ontological transformation, rendering it a central feature of the very constitution of meaning. The triangle of metaphor (Figure 2.2), on this account, models ontological transformation as a dialectical shift in the coordinates of meaning, while the 'spread out' triangle of obviation (Figure 2.3) models the inner complexity of its operation.

Crucially, Wagner's model of conceptualization as an interplay of invention and convention shows that the acts of conceptual invention that the ontological turn so prizes do not operate despite convention but *through* it. Accordingly, from a Wagnerian perspective, what distinguishes the ontological turn is not that it is somehow 'more' inventive than other anthropological approaches, since invention is involved in *all* meaning-production. The difference, as we also highlighted in this book's Introduction, lies in the deliberate and reflexive manner in which the ontological turn pursues conceptual invention as its prime and abiding task. As per Wagner's own account, anthropologists tend to instantiate a more general 'Western' propensity to treat invention as innate and convention as artificial: the world, as they describe it ethnographically, is as it is (viz. it is innate), and their job is to gain analytical control over it by bringing to bear on it a body of theoretical conventions (explanatory models, interpretive schemes, comparative procedures). While such operations are demonstrably inventive (theoretical conventions are elaborated, adjusted, critiqued, revised or even demolished through their exposure to the empirical contingencies with which they are supposed to 'deal'), this inventiveness is itself treated as an incidental by-product of the project of elaborating analytical conventions robust enough to have a purchase on the world, which they thus seek to bring under 'control'. The ontological turn, then, performs a figure/ground reversal of this image of human motivation. It attempts to imagine an anthropology that, like Daribi myth, takes conventions as read and treats them as the foil for deliberate acts of conceptual invention. Hence the constitutively reflexive and experimental character of its conceptualizations: like Daribi myth, the kind of analysis the ontological turn pursues depends on looking reflexively 'inwards' at conventional assumptions, obviating them experimentally, in order to produce novel conceptual effects and thus, as with myth again, to take anthropological thinking in new directions.

We may note that such an argument brings full circle a 'foundationalist' interpretation of Wagner as laying the theoretical cornerstones of the ontological turn, by showing how his analytical framework effectively encompasses *itself* and, by that virtue, allows us to articulate the contrast between the ontological turn and its putative adversaries. Still, there is a basic problem with treating Wagner as the theorist of (what eventually developed into) the ontological turn in this way. For if, as we argue in this book, the distinguishing feature of the ontological turn is its very refusal to take theoretical foundations for granted, then it is at least odd (and at most self-contradictory) to provide such an anti-foundationalist methodological injunction with theoretical foundations of its own. How reflexive exactly, and how experimental (let alone '*self*-experimental'), is a manner of anthropological conceptualization that is grounded in a prior conceptualization – a 'theory' – of what counts as reflexivity, experimentation and, indeed, conceptualization? The paradox is identical to the one encountered with Schneider earlier: if there the problem was with disowning the concept of culture in the name of culture, here the problem is with disowning the project of theory-building in the name of a theory one has built. Or, to borrow Wagner's own terms, the contradiction lies in treating Wagner's argument for anthropological invention as an attempt to establish an anthropological convention.

Whether Wagner himself falls into this trap is an open question. Certainly, when reading for instance *The Invention of Culture*, one often feels that his manner of developing (inventing!) his models comes close to traditional 'conventionalizing' theory building. This impression is borne out by the tendency, in this early period of his work, to premise his account on the more general 'symbolic anthropological' hypothesis that 'meaning' is the distinguishing feature of human phenomena (e.g. 1972: 4; 1977a). He is even sometimes drawn to some of the signature postures of conventional 'scientistic' anthropology, for example, when he seeks to ground his account of meaning in neuropsychological models of perception (e.g. 1986: 18), or when he offers generalizations as to the geographical distribution of his models (in *The Invention of Culture* he repeatedly claims that taking convention as innate and invention as artificial is something not only the Daribi, but all 'tribal', 'peasant' and 'religious' people do, while the inverse stance is a general characteristic of 'the West' – e.g. 1981: 74–5; cf. 1981: 7). Nevertheless, Wagner's deliberate (and increasingly cryptic) use of irony, humour and trickster-like wordplay over recent years (e.g. 2010) casts doubt on the degree to which such theoretical or 'ontological' affirmations are meant seriously as claims about what the world is 'really like' (viz. the representational semiotic of convention). At any rate, as with culture, one suspects that what counts as serious for Wagner and for the practitioners of ('serious') science may be two different things.

Be that as it may, the danger that this kind of theoretical foundationalism represents for the project of the ontological turn is indeed serious, as has been pointed out by a number of critics (Course 2010; Alberti et al. 2011; Laidlaw 2012; Scott 2013a; 2013b; Ricart 2014; cf. Holbraad 2012: 260–5; Pedersen 2012a). If the ontological turn is to be defined by its propensity reflexively to undo ontological presuppositions in the face of ethnographic contingency, then to premise it on a prior set of ontological presuppositions – a 'meta-ontology' (Heywood 2012) – about meanings, symbols, dialectics, obviation, revelation, or what-have-you, is at best half-measured and at worst downright contradictory. It is, in effect, to exclude the reflexivity of the ontological turn from its own scope (Holbraad 2013a), belying one of its defining characteristics, namely that its conceptual experimentations go 'all the way down'.

Heeding the criticism implies refusing to take Wagner's framework as a theoretical grounding. Instead, we should conceive of the analytical pyrotechnics that Wagner has been firing in and at the discipline decade after decade as an ongoing demonstration of the power of conceptual invention – an outcome of a form of thinking always on the move – a point to which we shall return in Chapters 5 and 6. We may even go as far as to embrace the manner in which Wagner's analytic allows us to articulate and sharpen some of the central tenets of our argument for the ontological turn, without excluding it willy-nilly from the very process of reflexive experimentation for which, perforce, it has opened up the way. Certainly, just such a possibility – of being thoroughly and permanently reflexive about one's manners of reflection – is one that has been taken up in increasingly radical ways since the time of Wagner's first writings on invention, perhaps more so than anyone by his fellow-travelling Melanesianist, Marilyn Strathern.

THREE

Relational Ethnography: Strathern's Comparisons and Scales

For it to make sense to say that Marilyn Strathern's work has any bearing on questions and problems of ontology, then surely the meaning and purchase of 'ontology' would need to differ from its conventional essentialist and absolutist (philosophical, metaphysical) connotations. After all, for many of her admirers from anthropology and other disciplines, her work captures just about everything that anthropology has got right (or, according to critics, wrong) since postmodernism. At issue is not only that *what* Strathern writes evolved in dialogue with wider intellectual-cum-political visions, critiques and projects formulated broadly between late 1970s and early 1990s, including feminism and the so-called crisis of representation; at issue is also *how* Strathern writes, and her hyper-reflexive attitude towards knowledge-making of all kinds. For, as Strathern herself puts it in a recent publication, 'What is true of what is observed is true also of the manner of observation' (2014a: 7). What could be more anti-essentialist and 'epistemological' (as opposed to 'ontological') than that?

In fact, we argue in this chapter, Strathern's work has played a decisive role in the development of anthropology's ontological turn, even if she herself remains 'awkwardly' (cf. Strathern 1987a) positioned with respect to it. Indeed, we are going to show, it is because of Strathern that the ontological turn is the natural heir to anthropology's postmodern self-critique of the 1980s. Without Strathern, none of the three trade-marks outlined in the Introduction to this book – reflexivity, conceptualization, experimentation – would have had the same critical edge. Had it not been for the distinct influence of Strathern on scholars such as Viveiros de Castro and others, the ontological turn would neither have been as reflexive nor as deliberately experimental about its own concepts and, crucially, their limits, as it is.

This, above all, is what distinguishes Strathern from Roy Wagner (another scholar without whom the ontological turn as we know it would not have existed). While firmly rooted in his theory of symbolic obviation, Strathern is nevertheless far more consistently oriented towards self-reflexivity than he is, as is evident in the explicit and systematic way in which she plays with the limits of descriptive language and analytical concepts. Certainly, the totalizing tendencies that we identified in our commentary on Wagner in the previous chapter are conspicuous by their absence in Strathern's work. So, while Wagner is cited in the front matter of her *The Gender of the Gift* (1988), so, tellingly, is feminist historian Jill Julian Matthews – an acknowledgement of Wagner's influence, coupled with an indication of the different intellectual traditions and political visions by which Strathern's work has been shaped and to which it has contributed. For whereas one would be hard pressed to describe Wagner as a feminist, Strathern's work has had a significant impact on feminist scholarship and vice versa, as illustrated, for instance, in Strathern's longstanding dialogue with Donna Haraway (e.g. Haraway 1989, 1991; Strathern 2004).

Later on in this chapter we shall be commenting on the significance of Strathern's feminism for understanding her place in the line of thinking that led up to the ontological turn. As we shall see, to understand her influence on the ontological turn, but also the distance she keeps from it, it is necessary to take into account Strathern's concerted attempt to cultivate a distinct analytics, in which 'incompletion' and 'hesitation' are treated not as obstacles to be overcome but as virtues in their own right. We begin, however, by addressing a third source of inspiration that is cited alongside Wagner and Matthews on the opening page of her magnum opus, namely Alfred Radcliffe-Brown. If, as we saw in the last chapter, Wagner's project takes off from the work of David Schneider and the American tradition of cultural anthropology more generally, then, as we show in the present chapter, Strathern's anthropological project is firmly anchored in the British tradition of social anthropology, and most notably the structural functionalist approach associated with Radcliffe-Brown.[1] Nowhere is this clearer than in Strathern's discussions of 'the relation' as an abiding anthropological matter of concern, which we shall take as our starting-point.

Relations Everywhere

It has been suggested that 'hardly anyone in social anthropology today claims to be a follower of Radcliffe-Brown' (Barnard 2000: 73). Nevertheless, while Strathern might not describe herself as a follower, she clearly sees her project as following on tracks set down by the

---

[1] Much as with Wagner and the American tradition of cultural anthropology (see Chapter 2), the suggestion that Marilyn Strathern is somehow 'representative' of the British social anthropological tradition must be qualified, since it bypasses what might be described as the non-Durkheimian (and therefore, if you like, 'non-French') part of the tradition promulgated and popularized by Malinowski and his students following the demise of Victorian anthropology (Stocking 1984: 106–91; Stocking 1986). Certainly, to highlight, as we are going to do below, the intellectual genealogy that can be traced from Durkheim through Radcliffe-Brown and Meyer Fortes to Marilyn Strathern downplays the significant interest in psychology that was such a prominent intellectual current in British anthropology not just during the Haddon-Rivers period, but also among the first generation of functionalists, including Malinowski (who found much inspiration in Victorian psychologists as well as Freud) and many of his students, several of whom had degrees in psychology (including, interestingly and perhaps surprisingly, Fortes, who experimented with psychoanalysis and also underwent and conducted therapy – Stocking 1986; H. Kuper 1984).

founder of structural-functionalism. Several of her writings contain lengthy discussions of Radcliffe-Brown's and his students' work (notably that of Meyer Fortes), and she has often reflected on her formative years as a Cambridge graduate student in the mid-1960s (see, for example, 1999, 2004, 2014a). What made structural-functionalism so seductive, Strathern explains in her own inaugural lecture as Professor at Cambridge, 'was the creative appropriation of The Relation, as at once the abstract construct and the concrete person to understand the totality of social life in terms of its own internal ordering... "social order" became simultaneously the description of society and the perceived means of its cohesion' (1995a: 10–12).[2]

Still, Strathern's concept of the relation differs qualitatively from that of the structural-functionalists in at least two ways. First, insofar it makes sense to say, as anthropology teachers often tell their students, that 'Radcliffe-Brown made British anthropology French' (Segal 1999: 132) by adopting the Durkheimian view that 'any explanation of a particular sociological phenomenon in terms of psychology, i.e. of processes of mental activity, is invalid' (Radcliffe-Brown 1958: 64), then Strathern has been equally sceptical of sociological and psychological explanations due to their tendency to reify 'society' and 'the individual' respectively. In this sense her perspective calls to mind one of her other sources of inspiration, namely R. R. Marett, who also 'criticize[d] any one-sided concentration on the group as sharply as he [did] the one-sided focus of British anthropologists on the individual, rejecting sociological determinism for its denial of individual free will' (Segal 1999: 137). But

---

[2] As she explains, '[W]ith their penchant for the concrete', Rivers and his student Radcliffe-Brown 'set the agenda for ... how to understand the totality of social life in terms of its own internal ordering' (1995a: 12). And, crucially, it was by virtue of a 'double emphasis ... on relations known to the observer as principles of social organisation and relations observed as interactions between persons' (1995a: 12) that 'British social anthropology remained closely tied to the conviction that at the heart of systems ... [was a] primary human ability to make relationships' (1995a: 14; see also 2014a).

whereas 'Marett advocate[d] a combination of ... the "social psychology" of the French with the "individual psychology" of the British' (1999: 137), Strathern's approach is *neither* sociological *nor* psychological. Indeed, what lies behind her most famous concept of 'the dividual' (Strathern 1988) is a desire to formulate a 'vocabulary that will allow us to talk about sociality in the singular as well as the plural' (1988: 13). As with all other anthropological concepts, this concept is the product of concrete fieldwork encounters or more technically 'ethnographic moments' (more on which follows), in Strathern's case her long-lasting work and interest in Highland Melanesia. Thus, Strathern argues in *The Gender of the Gift*,

[f]ar from being regarded as unique entities, Melanesian persons are as dividually as they are individually conceived. They contain a generalized sociality within. Indeed, persons are frequently constructed as the plural and composite side of the relationships that produced them ... This premise is particularly significant for the attention given to images of relations contained within the maternal body. By contrast, the kinds of collective actions that might be identified by an outsider observer in a male cult performance or group organization, involving numbers of persons, often present an image of unity ... Thus a group of men or a group of women will conceive of their individual members as replicating in singular form ('one man', 'one woman') what they have created in collective form ('one men's house', 'one matrilineage'). (1988: 13–14)

So, whereas for Radcliffe-Brown, 'social structure' was directly visible in the form of all the particular dyadic relations observable between two actors or groups of actors in a given society, for Strathern and other exponents of the so-called New Melanesian Ethnography[3] (e.g. Wagner

---

[3] Like the 'Melanesian model of sociality' (c.f. Chapter 1) and similar-sounding terms, 'the New Melanesian Ethnography' is a label that has been used by fellow Melanesianists commenting on and criticizing the distinct kind of analytical method that since the mid-1980s has been associated with Roy Wagner, Marilyn Strathern and several of their students as well as other Melanesian anthropologists influenced by them.

1977b; Weiner 1988; Battaglia 1990; Leach 2003; Reed 2004; Crook 2007), social relations are not just comprised of connections visible *between* persons, but also all the past and future connections presently invisible *within* them. Here, gender relations play a defining role. For the differences between the two dominant 'aesthetic forms' that are known by anthropologists as the two genders are dynamically replicated across multiple scales ranging from tribe, clan, household and person so that, at a given moment, one gender is rendered visible while the other one is 'eclipsed'.[4] Thus, as Strathern goes on to explain in *The Gender of the Gift*, Melanesian '[s]ocial life consists in a constant movement from one state to another, from one type of sociality to another, from a unity (manifested collectively or singly) to that unity split or paired with respect to another ... The singular person, then, regarded as a derivative of multiple identities, may be transformed into the dividual composed of distinct male and female elements' (1988: 15). So, to use a distinction invoked by several commentators on Strathern's work (Gell 1999; Jensen 2012), whereas the structural functionalists conceived of relations as 'external' in that they bridged an imagined space between social units and scales (individuals, households, clans, states, and so forth), for Strathern as for Wagner (1977b), relations are conceptualized as 'internal' in the sense that there is nothing that is *not* relational. Instead of relations between different units and scales, there are only relations between different kinds of relations – relations everywhere, indeed.[5]

---

[4] By 'eclipsing', Strathern refers particularly to a 'special feature of [the] concealment' (1988: 155) that takes place in gendered processes of production and exchange in Melanesia, namely the fact that, 'as in lunar eclipse, for the effects of [male agency] to be registered, there can be only partly concealment and not obliteration [of female labour]' (1988: 157). In more general terms, Strathern deploys this concept – which resembles but is not identical to Wagner's 'obviation' (see Chapter 2) – to describe various figure-ground reversals that she identifies in Melanesian and other ethnographic materials.

[5] As Strathern explains in a recent article, relations are the very 'membranes ... by which the heirs of the scientific revolution assemble, and dis-assemble, their knowledge' (2014a: 14). Thus, 'relations constantly appear as solutions to anthropologists' problems of description. Indeed, the more so-called "bounded" notions

Which brings us to the second sense in which Strathern's 'relation' differs fundamentally from that of Radcliffe-Brown and her other British structural functionalist forebears, namely the fact that, for her, social relations do not exist in an external world out there, but are an intrinsic and inevitable effect of the manner in which anthropological analysis is conducted ("relations nowhere", as it were). Ironically, even though Melanesia is sometimes described as a home of 'relational cultures' (Robbins 2004a: 292) 'the relation' does not feature prominently as an indigenous concern in Melanesia; in fact, it has been argued that some Melanesians have no concept of the relation at all (Crook 2007; see also Englund & Yarrow 2013). Far from being an indigenous term adopted by ethnographers in the hope of 'grasping the native's point of view' (Malinowski 1922), then, 'the relation' is a proxy by which 'scholars trained in the Western tradition ... through deliberate choice ... glimpse what "other" assumptions might look like ... through an internal dialogue within the confines of [their] own language' (Strathern 1988: 4). To claim, as anthropologists sometimes do, that this or that 'people' or 'society' (in Melanesia, Mongolia or what have you) is 'relational' would thus be a deeply un-Strathernian move, for it overlooks the fundamentally experimental and heuristic nature of her analytical method and descriptive language. If Strathern's anthropological project can be described as 'relationalist', then, it is only because her approach is to conceive of everything *as if* it was relationally configured. Indeed, the 'recursive' character of the ontological turn's reflexivity and its concept of relationality, which pertains to the collapse between objects and means of anthropological description as we explained in the Introduction, is the prime indicator of Strathern's abiding influence on this line of thinking.

of society and culture are held up to criticism, along with the systems and structures that were once their scaffold, the more relations, relationships, the relational, relationality, are evoked as prime movers (of sociality) in themselves ...' (2014a: 5).

Later in this chapter and the ones to follow, we shall explore in detail the consequences that this heuristic conception, and therefore inherently experimental and strategic application, of 'the relation' has for our understanding of anthropology as an intellectual endeavour. For now we wish to emphasize that, for Strathern in particular, anthropological *concepts always come from somewhere concrete*, and from there they cannot, and thus should not, be fully detached. In the case of 'the relation', this 'somewhere concrete' is English kinship terminology and imaginaries traceable back to the seventeenth century if not before (2014a: 6; 9–11), which, to be turned into anthropological concepts deployable in other ethnographic contexts, were 'stretched' first by the structural functionalists and then by subsequent generations of social anthropologists such as herself (1995a; 2014a). It is this capacity of 'the relation' in the English – but not necessarily other – language for simultaneously denoting something very concrete and very abstract that explains the interests and inclinations, as well as the success and endurance, of classic British social anthropology.

Note that this 'conceptual contingency' is different from – but not in contradiction with – the conceptual contingency that the ontological turn (and Strathern herself) has otherwise mostly focused on, namely the equally concrete nature of the relationship between the object of an ethnographic description and its anthropological analysis (in fact, one could say that the key argument of the present book is *also* about how anthropological concepts 'always come from somewhere concrete', namely the particular kind of concretion instantiated in turning ethnographic materials into anthropological concepts). As such, Strathern's reflections on the emergence of the concept of the relation in British social theory from the seventeenth century onwards might be said to add another side to that coin, namely the contingency of the origins of the analytical concepts that the anthropologist has at his or her disposal when confronted with ethnographic contingency. In other words, we seem here to be faced with a double ethnographic movement, one towards 'Melanesia' and one towards the history of British thought, which are played off against each other in one recursive and reflexive theoretical manoeuvre.

Anything can in principle be imagined as a relation, then, because relationality is intrinsic to the way in which certain general forms of Euro-American 'knowledge practices' (another favourite term of Strathern's) is made, including anthropological knowledge in particular. Strathern has often been criticized for her use of the term 'Euro-American' (see e.g. Josephides 1991; Keesing 1992; c.f. Carrier 1992b) and seemingly for good reason. For is not her frequent invocation of 'Euro-America', as indeed 'Melanesia', not a capitulation to the 'Us-Them anthropology' that anthropologists, not least from North America, have spent so much time trying to exorcize from the discipline's tainted past? Here, it is important to note that Strathern's use of the term 'Euro-Americans', as she explains it, 'refer[s] to the largely middle-class, North American/ Northern European discourse of public and professional life' (1995a: 42). 'Euro-American', in other words, for her is not a territorializing tag for a particular group of people (objectifying 'Euro-Americans' as a form of 'society' or 'culture' to be compared and contrasted with others) but rather a particular way of organizing knowledge that is characteristic of the activity of anthropology itself, as well as the broader milieu of knowledge-practices to which it contributes. In this sense the status of 'Euro-America' in Strathern's work is strictly analogous to that of its twin term, 'Melanesia', which, as Alfred Gell so cogently put it,

is not the actual nation states of Papua New Guinea, the Solomon Islands, Vanuatu, and so on, but a manner of speaking, or more precisely the site of certain problems of expression and understanding, peculiar to the cultural project of anthropology ... It has nothing intrinsic to do with the totally artificial and internally discontinuous ethnographic area that happens, for mostly rather bad reasons, to have been christened 'Melanesia' (Gell 1999: 34; see also Strathern 1988: 12–13; Reed 2004; Crook 2007).

In keeping with this book's emphasis on the fundamentally heuristic and experimental gist of the kind of anthropological analytics that has become known as the ontological turn, it should therefore always be kept in mind that 'Euro-America' (along with 'Melanesia', 'Amazonia' etc.) are intended as nothing other than tropes intrinsic to the conceptual economy and analytical language of which Strathern is one of the leading proponents.

It is clear, then, that the relations theorized by Strathern are fundamentally different from those studied by the structural-functionalists and indeed certain contemporary anthropologists, who seem to take relations to exist 'out there' as the basic building blocks of social, cultural, and even, natural worlds. This is why it is relatively straightforward to refute another frequent critique of Strathern's work, namely that it amounts to a 'theoretical metacosmology' where 'relational nondualism' is posited 'either directly or indirectly, as the way things really are' (Scott 2014: 32, 34; see also Heywood 2012; Venkatesan et al. 2013; Scott 2013a, 2013b, 2015b; c.f. Graeber 2015). After all, as Strathern makes clear, relations are neither to be understood as more real (let alone 'really real') in ontological terms, nor as superior in moral terms. On the contrary, for Strathern relation as concept is a limit point that anthropological thought cannot cross (although this still leaves us with the question as to what might happen to the concept of the relation *itself* when it is made subject to anthropological analytical experiments, as we shall be showing in Chapter 6). It is true that, when reading Strathern's work, or work inspired by her (e.g. Mosko 1985, 1992; Bird-David 1999; Myhre 2013), one is sometimes left with the impression that the relational method is the only way in which ethnographic descriptions can be made, or even that the world itself is comprised by relations in more metaphysical terms (see also Chapter 6). Yet, from Strathern's perspective, far from constituting the best possible method for performing an ethnographic description, 'relating' in the present sense emerges as the 'least bad' one, so to speak.

In sum, rather than representing an external social reality, relations here are to be understood as (British) anthropology's most successful 'technology of description' (cf. Pedersen 2012a). Relations are instruments of anthropological thought; they are what we do, not what we study (unless, of course, our object of analysis happens to be the history of English social and political theory, e.g. Strathern 1992a; 2014a, 2014b). For the same reason, it would amount to a serious misunderstanding of Strathern's project to describe her work as a 'relational ontology', as if her aim were to expound a full-blown metaphysics of her own. 'Relationism', as others have called it in the context of Melanesian ethnography (e.g. Robbins 2004a: 292), for Strathern is 'just' method.[6] Anthropologists do not study relations between the people they study, but the relations they need to 'invent' to study those people. For her, relations are not substantive in the sense of something that 'is', but contingent in the sense of something – an ethnographic moment – that happens in the encounter between anthropology and its subject matter.

But how is this actually done – what does it mean to study relations *as if* they are potentially anywhere? To explore this crucial question, we will now discuss in some detail the way Strathern brings her focus on relations to bear on the question of comparison, which for her is the quintessential form that 'the relation' takes in anthropological inquiry. Indeed, as we shall see, if the 'relational' is the necessary object of anthropological study, comparison is its only method. As we

---

[6] This is also why 'Strathern's project of anthropology, with its recurrent use of a contrast between Melanesia and the West, is not recourse to relativism' (Hirsch 2014: 42). To be true, a 'sense of relativism may emerge from the anthropologists' investment in relations, and from taking these relations across cultures' (Strathern 1995a: 25). But instead of mapping similarities and differences between contexts, '[o]ne must ... be prepared for the unpredictable, including different distributions of what people take as finite and what they take as infinite about their circumstances' (Strathen 1988: 249).

shall see, this focus on comparison (and the more technical notion of 'scale' which she introduces in this connection) takes us to the heart of what is at stake in her anthropology. It also takes us a long way towards realizing why her project cannot simply be described as 'epistemological' (as she generally prefers), but also as having inherently 'ontological' implications, if by that we understand the kind of concern with conceptual reflexivity and experimentation that lies at the heart of the ontological turn.

Comparison All the Way Down

Getting a handle on Strathern's concept of comparison is an exercise that instantiates the very problems it addresses (what is a comparison?). Comparison understood dually, as both a social activity that features in our ethnographies and as an anthropological method, permeates her works, so that discussing it becomes a comparative exercise in its own right – a comparison of comparisons. Mindful of the frustrations with reference to which she gauges the stakes involved in the task of comparison – the dizziments of disproportion, arbitrariness and assorted variables, levels, contexts, dimensions and so on running riot, we home in on the most explicit treatment of this theme in Strathern's work, namely her book *Partial Connections* (2004).

So what notion of comparison does Strathern have in mind in *Partial Connections*? The point is put at the book's outset by way of a comparison of commonplace strategies of comparison in anthropology, cast in terms of the concept of 'scale'. In line with modern 'Euro-American' metaphysical intuitions, Strathern argues, anthropologists imagine the world as consisting of many many things – an inordinately large field of data. So the most basic methodological question for anthropology (as for any other discipline) is how to bring this 'plural' data, as she calls it, under some kind of control. Put in very general terms, this must involve deciding which data go with each other and which does not. In this general sense all 'Euro-American' descriptive activity is indeed comparative, although there is also a sense in which the anthropological challenge of cross-cultural comparison is 'exemplary' (2004: xvi) in this respect, since the things compared – societies or cultures – are fields of phenomena that are defined precisely by the fact that their constituent elements somehow go together, the problem being to work out what these elements are and how they do or do not relate.[7]

In response to this challenge, Strathern shows, social scientists tend to plot their materials against different 'scales', understood as particular ways of 'switching from one perspective on a phenomenon to another, as anthropologists routinely do in the organization of their materials' (2004: xiv.) Such anthropological 'scalings' happen in two ways. For our purposes of exposition, here we may gloss the first one as quantitative scaling, since it involves switches in size, and corresponds to the conventional meaning of 'scale' as having to with numerical measurement and size. Like Gregory Bateson, for example, one might devote a book to a single ritual performed by a particular group of the Sepik River in Papua New Guinea (1958), or contrastingly one might devote four major volumes to the study of hundreds of myths from across the American continents, as Lévi-Strauss did (1969, 1979, 1990a, 1990b). The switches for which this kind of quantitative scaling allows depend

[7] It is important to stress that, as far as Strathern is concerned, comparison is an intrinsic feature of 'Euro-American' knowledge, but not necessarily 'Melanesian' also. On the contrary, she suggests that if 'we' do comparison, 'they' do something different, namely *division*. As she writes, 'Euro-American question[s] ... of difference [are] made manifest in comparison ... And I stress comparison rather than division in order to reserve the term division for a different mode of conceptualizing gender difference altogether ... [In] Euro-American social practices ...[a]s elsewhere the sexes are opposed and contrasted, their attributes seemingly divided off from one another; however ... such differences draw not on the kind of division found in the Melanesian ... cases but on a form of analogy that I have been calling comparison' (1995b: 43–44, 53).

on keeping the terms of comparison (its 'form') constant while shifting its scope (or 'content'), by scaling either 'down' to include more detail or 'up' to gain more purview. This mode of scaling, as Strathern writes

is made possible by a modelling of nature that regards the world as naturally composed of entities – a multiplicity of individuals or classes or relationships – whose characteristics are in turn regarded as only ever partially described by analytic schema ... The relativizing effect of knowing other perspectives exist gives the observer a constant sense that any one approach is only ever partial, that phenomena could be infinitely multiplied. (2004: xiv)

Yet, as she goes on to explain:

The interesting feature about switching scale is not that one can forever classify into greater or lesser groupings but that at every level complexity replicates itself in scale of detail. The 'same' order of information is repeated, eliciting equivalently complex conceptualization ... The amount of information remains, so to speak, despite an increase in the magnitude of detail. (2004: xvi)

This then suggests a second mode of scaling, which plays an equally prominent role in Strathern's account of comparison, namely one that maintains stable contents while shifting forms, and which may for our purposes here be glossed as qualitative – a more unconventional (metaphorical) use of the term 'scale'. Here viewpoints on a given body of data switch by changing the terms of reference one brings to bear upon it, as, for example, one does when one compares different cultures (or different elements 'within' one) from the point of view of economic arrangements, or ritual practices, or cosmological reckonings, and so on.[8]

---

[8] It goes without saying that, in ethnographic craftsmanship, any attempt at comparison will involve multiple combinations and mutual adjustments of both quantitative and qualitative scaling, and its success will depend on the skill – and the degree of reflexive awareness about its own undertaking as a knowledge-practice – with which this is done.

Now, these articulations of the act of comparison (themselves apparently forming a two-place qualitative scale for the comparison of different kinds of comparison) may seem already to describe one sense in which the connections on which comparisons rely might be 'partial', to use Strathern's terms. According to such an interpretation, scalings might be imagined as partial because, as finite human (viz. anthropological) acts, they can never contain the infinite plurality of the cosmos, and therefore always must leave a 'remainder', to use another favourite term of Strathern (2004: xxii). This, however, is not Strathern's point. Her concern is rather with the way in which infinity replicates itself *within* whatever scale purports to carve it. One may think that by changing one's viewpoint on one's material (e.g. scaling up to gain an overview of its general contours as opposed to scaling down to limit the amount of data considered, or shifting between different terms of reference altogether) one may reduce its complexity, but in doing so one soon realizes one is playing a zero-sum game. As indicated, for example, by the absurdity of saying that by virtue of its narrower ethnographic focus Bateson's *Naven* is a more simple read than Lévi-Strauss's *Naked Man*, or that Strathern's own oeuvre is less demanding for having homed in more on social interaction and gender relations than on religion and cosmology, the irony is that the potential for complexity remains constant no matter what the scale. To stick to the theological vocabulary of our gloss of Strathern's point, it is as if the notion that scaling can cut the cosmos down to size involves forgetting that infinity can be intensive as well as extensive, with angels dancing on the head of a pin just as well as in the ethers.

This insight is the basis of what Strathern calls a 'postplural perception of the world' (2004: xvi, see also 1992b), in which the notion that scales can act to carve finite, manageably simple parts out of an infinite, debilitatingly complex whole dissipates. If infinity goes both ways, both outward and inward, so that the scales that would purport to limit it end up acting as its conduits, then the very distinctions between plurality and singularity, whole and part, complexity and simplicity, as well as infinity and finitude, lose their sense. This is because the basic 'plural' assumption upon which each of these distinctions rests, namely that the world is made up of an infinite multiplicity of 'things' which may or may not relate to each other, vanishes also. If of every 'thing' (i.e. any object of anthropological attention) one can ask not only to what other things it relates (the pluralist project of comparison) but also of what other things it is composed, then the very metaphysic of 'many things' emerges as incoherent. Everything, one would conclude, is both more and less than itself. 'More' because what looks like a 'thing' in the plural metaphysic turns out, postplurally, to be composed of further things – infinity inward; and 'less' because at the same time it too contributes to the composition of further things – infinity outward.

This, then, raises the key question one needs to grapple with in order to understand Strathern's 'postplural' conception of comparison: what might comparison be in a world without 'things', understood as discrete, self-identical objects of anthropological attention? And if there are no things in that sense, then on what might anthropological comparisons even operate? On such an image, what would be, say, Melanesia and Britain, or the Western and the Eastern PNG Highlands, or, to use an example Strathern explores at length in *Partial Connections*, the different kinds of flutes (or methods of initiation, or modes of exchange, or whatever) that one might compare across them? In her own exposition Strathern presents a number of suggestive images: Donna Haraway's 'cyborgs' (1991), the mathematical concept of 'Cantor's dust', and, citing Roy Wagner (1991), the 'self-scaling fractal' with its capacity for 'not quite replications' (c.f. Green 2005) across multiple orders and levels of reality (Mosko 1985; Mosko & Damon 2005).

Here we want to stay with our own paradoxical formulation: *things that are what they are by virtue of being both more and less than themselves*. For the real 'virtue' of this paradox is that just as it renders incoherent the plural metaphysic of autonomous things, it serves as a coherent rendering of the postplural alternative. To be sure, *things* cannot be both more and less than themselves. 'More' and 'less' are comparatives after all, and it is hard to see the point of comparing something to itself, let alone of finding it different from, well, it. But this is just to say that the postplural alternative to 'things' is exactly that: *comparisons*. Stripped of the assumption that it must operate on something *other* than itself, that is exactly what a comparison would look like: something that is both more and less than *itself*. Which is to say that on a postplural rendition, the differences that plural comparisons measure 'between things' emerge as constitutive of those very same 'things', and can best be thought of as residing 'within' them. Accordingly, this would imply also that the plural distinction between things and the scales that measure them also collapses into itself: saying that differences are to be thought of as internal rather than external to comparisons is also to imply that there is no 'outside' point from which comparisons could be viewed, measured or, indeed, compared (see also Latour 2005 for a similar point). So comparisons are things that act as their own scales – things that scale and thus compare *themselves*, this being the postplural sense that Strathern elaborates.

This line of thinking takes us back to the concept of 'the relation'. That comparisons are scale-shifting relations in Strathern's sense goes without saying. 'The Relation', as she puts it, '[being] itself neither large or small, can cross scales' (1995a: 17). Yet, we now suggest, conceiving of Strathern's entire theoretical universe as made up of 'comparisons' adds something new to it. In particular, a focus on the notion of comparison in her work redresses a potential source of dissatisfaction with her concept of the relation, namely the apparently inordinate malleability of the heuristics it produces – the virtue that Strathern's relations appear to make of a complexity that can 'run riot', to recall one of her own formulations. As we shall see, this becomes especially clear when the contrast between 'plural' and 'postplural' forms of comparison are articulated in starker and more explicit terms than she does herself. In fact, it may be because Strathern does not offer an explicit account of this contrast that her project has sometimes been mistaken for a kind of postmodern-sounding relativism.

Consider the following contrast of images Strathern offers in *Partial Connections*:

[The map] implies the existence of certain points or areas, like so many villages or fields seen from the air, that will remain identifiable however much their features are replotted; all that changes is the perspective of the observer. [The tree] implies some kind of closure that defines a system of concepts and their potential transformation from within, insofar as only particular trajectories are 'genetically' possible from the principles one starts with. (2004: xvii)

The images of the map and the tree correspond to what we have called 'quantitative' and 'qualitative' scales of comparison. Scaling up and down to alter a form's scope over content corresponds directly to what one means by 'scale' when referring to a map: the proportion that holds between a territory (content) and its depiction (form). Analogously, qualitative switches from one form of comparison to another (e.g. focusing on economic as opposed to religious dimensions of a given set of data) involve the assumption that each of these forms is related to the others in terms of the lateral and vertical relations that make up a genealogical tree. For example, while one might imagine economic and religious scales to belong to the same 'generation', like siblings, one might posit the scale of the 'social' to contain them both, like a parent. The two images are themselves laterally related (on a tree they would be siblings) inasmuch as they both make the control of data possible by virtue, in Strathern's words, of the 'constancies' they imply (2004: xvi-xvii).

Both images are to be contrasted to the imagery with which Strathern depicts postplural comparisons – postmodern cyborgs, fractals and so on. While Strathern puts these metaphoric depictions to all sorts of uses in her argument – thus displaying, one might say, the sheer malleability of the concept of comparison itself – one also gets the impression that a notion of a *lack* of control or, put more positively, an inordinacy of potential, acts as their cumulative effect. So, for example, if maps and trees rely on the constancies of identity and closure to contrive a sense of control over data, the cyborg suggests an image of inconstancy, or even incontinence: it 'observes no scale', being a 'circuit of connections that joins parts that cannot be compared insofar as they are not isomorphic with one another' (2004: 54). Indeed, the image of the fractal itself, with its 'not-quite replication' (ibid.: xx) that generates a 'proliferation of forms' (ibid.: xxi) inwards and outwards all the way, may produce in the reader a sense of asphyxia as well as one of beauty, vertigo as well as wonderment. Equally, it may provoke a typical quip made against 'postmodernists' at the time *Partial Connections* was originally written, namely that of anything-goes 'flatness'. The impression could be borne out by the punch line 'postplural realization' that gives the book its name: 'The relativizing effect of multiple perspectives will make everything seem partial; the recurrence of similar propositions and bits of information will make everything seem connected' (2004: xx).

Yet, we would suggest that something more novel and interesting lies at the heart of Strathern's characterization of postplural comparison – an extra dimension to her thinking on which she never comments explicitly in *Partial Connections* or elsewhere in her work, but which is nevertheless present in the manner in which she conducts her comparisons. This concerns the peculiar role that something akin to 'abstraction' plays in Strathern's analytics – although we wish to emphasize that what is at stake here is something different than the logical operations one ordinarily associates with that term. As we shall see, this point about abstraction also connects her work explicitly to the development of the ontological turn's investment in experimental forms of reflexivity and conceptualization.

The closest Strathern comes to an explicit statement of her concern with abstraction in *Partial Connections* is, tellingly, not as part of characterizing her own concept of comparison, but in the course of her discussion of other anthropologists' attempts to provide an integrated frame for comparing societies from the Highlands region of Papua New Guinea with reference to a theme they are meant to have in common, namely the association of the use of bamboo flutes with male power (e.g. Hays 1986). The problem with such cross-cultural comparisons, she argues, is that while they do pick out significant ethnographic and historical connections, they also, necessarily, involve a slippage of levels. From where, one may ask, do they draw the features of the common theme whose variations they wish to chart? If, for example, in some cases flutes are focal to male initiation while in others less so or not at all, or in some cases the flutes themselves are conceived as male and in others as female or as both, while elsewhere bamboo flutes are absent altogether, then from which of these cases does the putatively 'common' notion that flutes are an important element of male power draw its strength? As Strathern herself writes, 'The difficulty with this comparison is that our supposed common regional culture is composed of the very features which are the object of study, the "meanings" people give to these instruments, the analogies they set up ... [T]he common cultural core, the themes common to the variations, is not a context or level independent of local usage' (2004: 73).

At issue here is the familiar anthropological charge of essentialism: mistaking ethnographic categories for analytical ones. Yet, Strathern's remedy is anything but the familiar reminder or tautology of saying that all categories are by definition cultural, and that therefore the modern chimera of a culturally neutral analytical language for comparison should be replaced by the wiser proposal for a culturally laden dialogue, tutored by the anthropologist's own cultural and political reflexivity – that is, the crisis-of-representation move (see below). Rather than treating the slippages of levels that essentialism entails as grounds for its rejection, she makes a *virtue* of them.[9]

From a plural starting-point, slipping from putatively neutral scales for comparison to culturally laden objects of comparison (viz. essentialism) is indeed a problem. But from the postplural position, that is precisely what comparison is: the 'unwarranted' melding together of what the plural rendering posits as 'scales' and their 'objects' (things that scale themselves or equally, to complete the image, scales that 'thing' themselves). Whereas, on the plural imaginary, comparisons occur between different things, then, postplurally, they also take place within things, precisely because the postplural move is to treat everything – each and every thing in the world – *as* a comparison. As we now show, recognizing these implications of Strathern's comparative analytics allows one to arrive at a stronger characterization of her work than the rather bland brand 'relational' – its extra, and if you like ontological dimension.

Postplural Abstraction

In plural terms, Strathern's comparisons evince a failure of abstraction. As a 'scale' for comparing Highlands societies, flutes and male power are not abstract enough, because they do not constitute a 'level' of analysis that is consistently of a different logical order from the cultural 'contexts'

---

[9] As Strathern writes in *The Gender of the Gift*, 'My account makes explicit one common implicit practice: extending out from some core study certain problems that become – in the form derived from the core study – a general axis of comparative classification ... What becomes objectionable in much comparative analysis is the decentering of the initial correlation, as though it somehow belonged between or across several societies and was not in the first place generated by one of them' (1988: 45–6; see also Schlecker and Hirsch 2001; Morita 2013; Pedersen & Nielsen 2013; Englund & Yarrow 2013).

that are meant to be compared. Indeed abstraction of just this kind is integral to the *plural* notion of comparison: for scales to be able to measure things they have to be more abstract than them. Now, clearly the distinction between abstract scales and concrete things cannot survive the transition to thinking of comparison postplurally unscathed, the whole point being that in such a transition the very distinction between scales and things is obliterated. Nevertheless, we argue, something of the distinction between the abstract and the concrete does survive – a residue or 'remainder', in Strathern's own terms. To see this we may consider once again the plural operations she presents.

How is conventional, plural abstraction *supposed* to work? Consider the verb: 'to abstract' something involves isolating from it one of its predicates. Take, say, a dog and isolate from it its quality of being a 'quadruped'. Or, to recall Strathern's own example, take the flutes PNG Highlanders use and isolate the quality of being 'associated with male power'. As we have seen in relation to Strathern's comments on the role of scale, such acts of isolation afford a battery of techniques that are supposed to help bring data under control for purposes of comparison – not least, quantitative scoping by analogy to maps and qualitative ordering by analogy to genealogical trees. To take our own rudimentary example, we assume that abstracting from a dog the quality of being a quadruped allows us to make analogies between it and a cat, or to study it from the point of view of its locomotion, contrasting it perhaps to other quadrupeds whose legs are otherwise different, or relating it evolutionarily to bipeds, or placing it within in the class of mammals, and so on. Abstraction increases the agility of comparison, one might say.

This example shows how central Strathern's observation regarding the paradoxical notion of control – the idea that no matter what the scale the degree of complexity stays constant – is to this way of thinking of abstraction. Just as 'isolating' a particular predicate would suggest a reduction of complexity (a dog is so many things other than a quadruped), so the very same act gives rise to new orders of complexity (new analogies, classifications and other such orders of relation). Thinking of this paradox in terms of abstraction, however, serves to reveal further features of the constancy of complexity that make it seem less than a 'riot'. Two hold particular interest.

First, the idea that abstraction entails 'isolating' predicates of objects allows us to emphasize one aspect that Strathern's characterization tends to leave mute, namely the idea that what she calls scales can be said to *originate* in the things they serve to compare. Indeed, the manner of the origination is just as interesting as the fact. While the thought of comparing things 'in terms of' or 'with reference to' scales conjures a notion of application (as, one might say, a rule applies to instances), the obverse thought of originating abstractions (scales) from more 'concrete' objects brings to mind a notion of extraction: to isolate a predicate is to *cut it away* from the denser mass in which it is initially embroiled, that is, what looks like 'the thing'. To use the sculptor's figure/ground reversal, comparison on this account involves cutting away the mass to make the abstraction appear.[10]

This brings us to a second characteristic of abstraction, which has to do with notions of removal and distance. We have already seen that such notions are foundational to the ontological assumptions of pluralist

[10] This 'creative cutting' (Pedersen 2014) is integral to the fractal imagery of 'Cantor's dust', in which scalar effects are replicated by the creation of intermittencies and gaps (Strathern 2004: xxii–xxiii), and, indeed, to the entire organization of *Partial Connections* as a text and an argument. In the foreword to the updated edition, Strathern thus explains how it was composed with the intention that 'every section is a cut, a lacuna: one can see similar themes on either side, but they are not added to one another' (2004: xxvii). Note the characteristic sense of 'cutting' here, which is used not in the sense of reducing complexity (its conventional, 'plural' sense of making a generalization), but as a particular conduit for (scale of) complexity itself: '*Partial Connections* was an attempt to act out, or deliberately fabricate, a non-linear progression of argumentative points as the basis for description ... Rather than inadvertent or unforeseen – and thus tragic or pitiable – partitionings that conjured loss of a whole, I wanted to experiment with the apportioning of "size" in a deliberate manner. The strategy was to stop the flow of information or argument, and thus "cut" it' (2004: xxix).

comparison, since 'distance' is precisely what is here imagined to sep-arate not only things from each other, but also things from the scales that are brought to bear on them. It is just such distances that images of maps and trees conjure – scaling up or down on an axis of proximity and distance, or branches and stems that are related vertically and horizon-tally by degrees of inclusion and exclusion. For scales to offer a vantage point from which things can be compared they have to be posited as being separate from them – perspective implies distance and indeed size. Thinking of comparison in the key of abstraction, however, foregrounds movement as a condition for both. If abstraction involves cutting predi-cates away from the things to which they belong, the distance it achieves can be conceived as the result of an act of *removal* – a trajectory that *cuts open* a gap.

Two thoughts about abstraction, then, are embedded in Strathern's account of the pluralist metaphysic of comparison: the notion that the things can scope their own comparisons by being cut (multiplying their comparative potential, so to speak, by being divided) and the notion that this involves a trajectory of movement. Both of these features carry over to Strathern's characterization of 'partial connections' – i.e. her account of what comparisons involve when one shifts to a postplural metaphysic, in which the distinction between scales and things is collapsed. Indeed, we would argue that they can be used as the basis for a suitably altered conceptualization of the notion of abstraction itself, one that we believe goes to the very heart of Strathern's thinking on comparison.

We call 'postplural abstraction' this alternative form of abstraction, which emerges when the binary distinctions between the specific and the general, as well as the concrete and the abstract, are themselves over-come. We have already seen how the postplural move involves rendering internal to things the differences that scales of comparison would find between them, thus turning things into self-comparisons, or could we say self-relations. Clearly the ordinary associations of abstraction with hierarchically ordered 'levels' separated from each other by degrees of distance (the images of maps and trees) have no place here. Nor does the corollary of this way of thinking, according to which abstractions represent things in more 'general' terms – as the concept of quadruped stands to any 'particular' dog. Indeed, one way of characterizing postplural abstractions would be to say that they are what abstractions become when they are no longer thought of as generalizations, i.e. as concepts that group together in their 'extension' things that share a particular feature.

Instead, postplural abstraction is what happens to abstraction when it turns *intensive*, in Deleuzian terms (De Landa 2002; c.f. Deleuze 1994). Postplural abstraction, then, refers to the capacity for things-cum-comparisons to transform *themselves* in certain ways. Considering our rudimentary example once again, postplural abstraction is what happens to a dog when it is considered *as* a quadruped. To think of a dog as a quadruped does not involve positing a relationship between two elements – a dog (deemed as a 'particular') that 'instantiates', as philosophers say, the concept of quadrupedness (deemed, in this sense, as a 'universal'). After all, the distinction between particular things like dogs and universal concepts like quadrupedness is exactly the distinction from which a postplural analytics moves us away – just a version, surely, of the distinction between concrete things and abstract scales which renders the world a 'plural' place. So, to consider a dog as a quadruped, on the postplural image of abstraction, is just to turn it (or 'scale it') into something different, namely, that thing-cum-scale that one would want to hyphenate as 'dog-as-quadruped'. This new 'third' element is a self-comparison in just the sense outlined earlier: it is 'more than itself' because, *qua* dog-as-quadruped, it is a full-blown dog; but it also 'less than itself' because, again *qua* dog-as-quadruped, it is merely a quadruped that has been postplurally, as opposed to plurally, abstracted.

To bring out the peculiar 'sharpness' of postplural comparison, we may supplement the range of images that Strathern uses to convey her

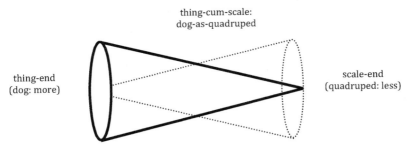

thing-cum-scale:
dog-as-quadruped

thing-end
(dog: more)

scale-end
(quadruped: less)

Figure 3.1 Postplural abstraction.

notion of comparison ('the fractal', 'the cyborg' and so on) with what one could claim is their most rudimentary form – the shape of a cone laid on its side (see Figure 3.1).

Imagining anthropological comparison in this way serves, first of all, to illustrate the crucial differences between postplural abstraction and its plural counterpart, which Strathern depicts with the twin images of the tree and the map. Conventional comparisons posit distances that separate both things from one another and things from the increasingly abstract generalizations in whose 'extensions' they are included. Moreover, the relationship between things and their generalizations is irreducibly hierarchical, since what makes generalizations suitable as scales for comparing things is that they are more abstract than the things compared. Postplural abstractions have neither of these characteristics. What in 'plural' abstraction look like extensive gaps 'between' things (and between things and scales) in the postplural mode figure as intensive differentiations 'within' them, indicated in Figure 3.1 by the asymmetrical proportions of the two 'ends' of the postplural abstraction – the broad 'thing'-like end and the sharp 'scale'-like one. Furthermore, the lack of a straight vertical axis indicates that hierarchy is absent here. Laid on its side, the hierarchical dimension that in plural terms marks the distances between things and scales dissipates into the internal self-differentiation within them.

This correspondence between the 'verticalization' of ordinary abstraction and the lateral self-differentiation of postplural abstraction helps to explain why Strathernian comparisons are conceptually sharper than just 'relations'. After all, it is the loss of the ordering principles that hierarchies of abstraction (and their corollaries in terms of inclusion and exclusion, connection and disconnection, similarity and difference, and so forth), that critics of the postmodernist penchant for profligate relations lament. So the formal correspondence between hierarchy and self-differentiation raises the prospect of retaining, if not a set of ordering principles as such, then at least a principle of a (no doubt new) kind of order, that indicate why Strathern's postplural universe is more than a magma of relations (cf. Scott 2007: 24–32). Might the asymmetry of self-differentiation do for the postplural universe what the symmetry of hierarchy does for the plural one? To see that this is so, we return to the question of 'cutting' and 'removal' that we introduced earlier.

Plural abstraction, as we explained earlier, involves the idea that scales of comparison are derived from the things they compare in two moves. First, deriving predicates (e.g. quadruped) from things (e.g. dog) by 'cutting' away from them the denser, 'thingy' mass in which they are initially embedded. And second, creating a distance between them and the mass from which they are extracted by placing them at a different level of abstraction, thus creating a gap between predicate and thing by a step of 'removal'. Each of these moves has a direct equivalent in postplural abstraction. First, when the difference between thing and scale is 'internalized' in the abstraction, the latter is still derived from the former. Only now, the sculptor's figure-ground reversal (viz. cutting the mass of the thing 'away' to make the abstract predicate appear) is reversed back: the mass of the thing is retained, but chiselled into a sharper, more elongated shape – still the same mass, that is, but 'less' than itself at its 'scale-like' end (to visualize this, imagine how the cone of Figure 3.1 might be sculpted out of the mass of a right circular cylinder). Second, while this 'internal derivation' of the scale from the thing does not involve opening up an (external) distance between the two, it does still turn on an act of removal, namely the 'internal' removal of the self-transforming proportions of the cone, as one moves from its broader end to its sharper one (again, to visualize this, imagine the motion of the sculptor's gouge as it cuts into a cylindrical mass to give it the shape of a cone). So what in the plural image were extensive *distances 'between things'* now become intensive *transformations 'within things'*, that may in this capacity be conceived as 'internal motions' – motions that are perhaps not unlike the ones classicists appreciate in the 'rhythms' of ancient columns.

Strathern's postplural universe comprised by what we have called postplural abstractions, then, presents an image that arguably comes close to what Lévi-Strauss had in mind when he spoke of the 'science of the concrete' (1966), provided we remain clear on the oxymoronic character of that phrase, where 'science' is meant to have connotations, precisely, of abstraction (as opposed to, say, objectivity). Certainly, as several commentators on Strathern's work have pointed out (Gell 1999; Viveiros de Castro & Goldman 2008; Hirsch 2014), Strathern's anthropological project has more in common with Lévi-Strauss than is sometimes assumed, especially when it is recognized that the 'structuralism' that serves as an inspiration here is not the impoverished 'scientistic' version promulgated by some cognitivist anthropologists (Sperber 1985), but the 'mythical' image of a fully relational universe composed by self-differentiating transformations espoused by Lévi-Strauss at the late-career peak of his theoretical vision – a point to which we shall return, also in the next chapter. For the time being, let us end our discussion of Strathernian comparison by observing that, just as Lévi-Strauss forcefully argued for the sophistication of 'savage thought', Strathern's postplural abstractions are in no way inferior to 'plural' abstractions when it comes to the sheer agility and scope of the comparisons they furnish. Only now this agility is no longer a matter of adopting different purviews onto things from the vantage points that more abstract scales afford (e.g. grouping cats and dogs together on grounds of their common quadrupedness and then contrasting them, say, from the viewpoint of their locomotion, possibly followed by some sort of synthesis). Rather, the potential for comparison is enhanced by the capacities that what a plural metaphysic would call 'things' (e.g. the dog) have to be transformed by being 'cut' in certain ways, 'sharpened' or 'contorted' so as to have particular aspects of themselves revealed (e.g. the dog-as-quadruped). And crucially, the effect of such transformations is that of providing, not a point of more general vantage and overview, but rather one of *further departure*. As thing-like (and scale-like) as the dog from which it was derived, the dog-as-quadruped presents further possibilities for comparative transformation in a whole spectrum of directions – including cats, locomotion, mammals and so on. Here, comparison does no longer occur with reference to a higher level of abstraction and generalization by reducing the individual complexities of and thus the differences *between* the objects of study, as in more conventional models of comparison. Rather, comparison occurs at the same order of reality and concretion as the object of study by unearthing differentiations *within* it.

Thus, the objective of Strathernian comparisons is not to generalize over different cases of social reality, but to *re-complexify* this reality on the scale of the analysis itself via controlled experiments in ethnographic comparison and conceptualization – a sort of 'savage anthropology' in Lévi-Strauss's sense, where the anthropologist emerges as a *bricoleur* of concepts not imposed on, but extracted from, the ethnographic moments studied. One may, then, think of Strathern as a conceptual sculptor, who works by eliciting certain dynamics and potentials present within things into intensified versions of these things themselves, not unlike an artist probing and sensing her way through the bundle of forces that the affordances of her materials enables or even compels her to release. Stretching the image, this is done by unearthing vantages within the 'dark side' of things, a sort of ethnographic worm holes that allow the anthropologist to plummet into hidden layers of relational potentiality latent within things so as to unearth new scalings or transformations of them – a sort of postplural or could one say hyper-things that are, in a paradoxical sense, 'more of the thing' than the thing itself.

This conclusion, itself intensely abstract, goes to the heart of one of the most compelling characteristics of Strathern's way of conducting comparisons, namely its sheer originality. While it goes without saying that one hardly needs to be a Strathernian to be original, we would argue that the work of postplural abstraction is inherently oriented towards originality of a distinctly anthropological kind. 'De-familiarization', as Strathern herself puts it, 'is crucial. But one cannot de-familiarize the whole world all at once; one has to proceed from the side, literally from the eccentric, in order to make the most obvious of questions seem not so obvious after all' (1995a: 61). Indeed, as we have seen, one way to express the contrast between plural and postplural abstraction is to say that while the former involves an 'upward' (as in the tree) or 'outward' (as in the map) move from the particular to the general, the latter moves sideways eccentrically, from particular(-cum-universal) to particular(-cum-universal), by means, as we saw, of eliciting dormant capacities for self-transformation within things. Here, comparison is no longer a matter of identifying general scales that may act as 'common denominators' for relating things (as 'quadruped' may relate cats and dogs). Rather comparison is oriented towards revealing 'uncommon denominators', in the form of peculiar and specific capacities for transformation that all things-cum-scales hold contingently within themselves. This is what Strathern calls 'scaling' and which we have here called 'postplural abstraction': the forging of qualitative equivalences via the identification of novel ethnographic analogies between seemingly disparate ethnographic materials.

Might the uniqueness of Strathern's analytical method boil down to this ability, evident in her thinking and writing, to avoid making the most obvious connections between different bodies of ethnographic materials – characteristically, in her case, 'Melanesian' and 'Western' – by 'cutting open' the least obvious (and thus most original) lines of comparison?

Certainly, as Street and Copeman write in a review of Strathern's work, her 'precise choreographed control of her materials seems *intentionally oriented* to the generation of surprise [through] the almost excessive "pruning" of her ethnographic images in order to excise familiar similarities and to reveal unintuitive analogies. In other words, her control of her materials appears to go as far as to actively seek unpredictable effects' (Street& Copeman 2014: 32). But precisely how does Strathern sharpen her concepts and recalibrate her analytical scales to establish the conditions of possibility for surprise? – which technologies of description does she use to generate her unlikely comparisons? Strathern's relationship to the 'crisis of representation' that swept across anthropology in the 1980s offers a useful starting point for exploring this.

Deep Hesitation

As already discussed in the Introduction to this book, the postmodern 'crisis' that anthropology purportedly underwent in the 1980s was, essentially, an attack of disciplinary self-consciousness that took the form of an intense concern with the question of anthropological reflexivity as well as styles of ethnographic writing. Imagining earlier generations of anthropologists as having ignored in the name of positivist objectivity the irreducible influence of their own personal, cultural, and political biases on their research, the idea was to re-invent anthropology by making these hitherto tacit influences and genres explicit. After all, it was recognized, anthropology is itself a socio-cultural practice, and hence belongs to the same order of phenomena that it purports to study. What was called for, therefore, was an anthropology imbued with a double vision: one eye on the object of inquiry, the other on the inquirer. And what made this move a 'crisis of representation' was that it had the potential to bring down the entire project of modern anthropology, understood as the endeavour to arrive at accurate representations of social and cultural phenomena which could provide the basis for theoretical generalizations: no more modernist naïveté, was the message. But for its detractors (e.g. Spencer 1989; Sangren 2007), the real crisis resided in the remedy of reflexivity itself. For if the conditions of possibility of anthropological knowledge are to become part of the object of knowledge, then what are the conditions of possibility of *that*? Which is just a formal way of expressing the habitual – but in retrospect not always fully justified – quip against the reflexive and postmodern 'turn' in US anthropology in the 1980s: 'navel gazing'.

In the preface to *Partial Connections*, Strathern recalls how she structured the book 'in response to … changing anthropological approaches to writing and representation in ethnography' (2004: xiiv), by which she in all likelihood refers to *Writing Culture* (Clifford & Marcus 1986), *Anthropology as Cultural Critique* (Marcus & Fisher 1986) and other famous crisis of representation texts. Nevertheless, while she is clearly sympathetic towards the critical and experimental aspirations of Rabinow, Clifford, Tylor and other scholars associated with this period, she also makes it perfectly clear, both in *Partial Connections* and elsewhere in her work, that they did not go far enough in their attempt to formulate a 'conscious theoretical framework challeng[ing] existing theoretical frameworks' (1987a: 277).[11] In particular, Strathern's reservation appears to be that the cultural reflexivists failed to fully identity the limits of representational language – not, crucially, in order to transcend these limits (an aspiration towards which Strathern is sceptical), but with a

[11] This is why Tyler's vision of an 'evocative ethnography' (Tyler 1987) falls short. As much as he was 'impatien[t] with the idea of representation' (Strathern 2004: 14) and celebrated 'incompleteness' as the baseline condition, his 'cognitive utopia' of a 'unifying pastiche' (2004: 16) still rested upon a very modernist longing for a 'return to an idea of integration' (2004:14) personified in the 'transcendental' (2004: 22) figure of 'the cosmopolitan' (2004: 23). Instead of this well-known critical 'inter-subjective' reflexivity, where the supposedly transparent subjectivity of the ethnographer's self is made into an object for introspection (Rabinow 1977), Strathern offers what might be called an 'intra-objective' alternative, where the 'objectivity' of the anthropological self is rendered into an ever less stable – and ever less transparent – 'scaling' of itself.

view to taking representation (and its crisis) so seriously that one is able to push and extend its boundaries from within. While Strathern does not say this directly, one of the things that she appears to find lacking in the work of the postmodern anthropologists is the fact that, as much as they experimented with writing and authority, they developed their critique, and alternatives, within the very representational logic from which they were so eager to escape. In other words, the problem was not that they were too reflexive, as some of their critics have objected (see e.g. Sangren 1988), but that their reflexivity did not go far enough.

In contrast to the crisis of representation literature, we suggest, Strathern in her writings avoids or deflects the charge of navel gazing. Taking to its ultimate consequence the postmodern injunction to treat the self both as an object and a subject of scrutiny, we argue, Strathern effectively comes out on its other side. At whatever scale one might choose to recognize it, the 'self' is eliminated as the subject of anthropological analysis and thus features only as its object. Unwilling to partake in anthropological self-therapy, Strathern's texts enunciate a self that is perpetually obviated in a process of what might be dubbed 'extrospection', which, to borrow a description of Melanesian persons from *Partial Connections*, allows for 'the centres of others [to] become centres for [itself]' (2004: 117).[12]

But precisely how does one do this – where to find the intellectual, political and aesthetic inspiration and ammunition to construct this 'third way of personifying the ethnographic experience, to draw a figure who seems to be *more than* one person, indeed more than a person?' (2004: 27). Here, we need to consider what Strathern has famously

---

[12] Several commentators have reflected on the manner in which Strathern deliberately seems to absent herself from her texts. In her review of *Gender of the Gift*, Margaret Jolly thus notes that '[j]ust as the individual is expunged in the analyses of Melanesian personhood, so the author eludes us' (1992: 146). 'The effect', adds Tony Crook, 'is that the author appears to have disguised herself in order to let the methods [for description and analysis] first be seen for themselves; indeed, it is as if the methods were making exchanges amongst themselves' (2007: 75).

described as the 'awkward' (Strathern 1987a) relationship between feminism and anthropology as intellectual and political projects. Strathern's feminist inspirations have been well accounted for both by herself (1987a, 1988) and by others (Stacey 1988; Moore 1988; Lebner 2016), so there is little need to go into them in detail. Suffice to say that, arguably, the most important lesson that she took from feminism, and which irreversibly changed her anthropological project, was neither the focus on women per se (even if it certainly was that too), nor the relentlessly critical and sceptical stance towards the powers that be and their conventions (though feminism's revolutionary and emancipatory potential has clearly mattered a great deal to her and her thinking, c.f. 1988: 22–40; 1987a). Above all, at least when viewed from our particular perspective, what Strathern took from feminism was *non-completeness* as an analytical attitude, and, indeed, a virtue. As she writes, 'the idea of an incomplete project suggests that completion might be possible; feminist debate is a radical one to the extent that it must share with other radicalisms the premise that completion is undesirable' (1988: 22). Or as a Donna Haraway memorably put it in the final sentence of her famous manifesto, 'I would rather be a cyborg than a goddess' (Haraway 1987: 316).

But what are the ramifications of this anti-perfectionist injunction behind feminist anthropology (and anthropological feminism) – how to go about deferring completeness in practice? The answer may be found in what in the aforementioned paper she refers to as the 'doorstep hesitation' (as opposed to embracement or antagonism) between feminism and anthropology – a sort of creative tension, or productive misunderstanding (Tsing 2005), where each 'mocks the other, because each so nearly achieves what the other aims for as an ideal relation with the world' (Strathern 1987a: 286). Could this be the method by which one can practice non-completeness as a scholarly, political and ethical goal? Awkwardness and eccentricity, that is, as a bulwark against too fast and too easy totalizations, – after all, is that not what hesitation is: non-completeness temporalized as conduct? Indeed, we would go as far as suggesting, 'hesitation' comes as close to a description of Strathern's intellectual sensibility as one can get, and especially so in view of how carefully, systematically and pervasively she adopts it as an analytical attitude, including, tellingly, in her comparisons between feminism and anthropology and, as we shall see, her stance towards the turn to ontology.[13] 'Deep hesitation', we are going to show, to Strathern is an indispensable tool for the forging of unlikely postplural comparisons, for it forces the anthropologist to insert a productive pause between her ethnography and her intuitions.

To make this point, we need to return for a final time to the question of what anthropology's crisis of representation might have looked like had it not developed within the pluralist metaphysics of the 'writing culture' approach but had instead unfolded within the postplural analytics developed by Strathern. Here, it is useful to consider the more or less veiled critique she has made of various attempts during anthropology's post-writing-culture phase to 'modernize' fieldwork and attune it to the study of 'the global' (Hannerz 1993; Marcus 1995; Appadurai 1996; Gupta & Ferguson 1997; Olwig & Hastrup 1997; Hylland-Eriksen 2007). For example, the problem with George Marcus's concept of multi-sited fieldwork (1993) is the pluralist assumption behind the notion that the supposedly limited scale of 'the local' is overcome by conducting fieldwork in several different places. By 'following the people', or 'following the thing', the assumption seems to be, the ethnographer of 'global assemblages' or 'actor-networks' becomes imbued with the capacity to adopt a trans-local vantage from which disparate phenomena and perspectives may be brought together into a single, if fragmentary and non-totalizing narrative. Yet, as Strathern shows in her critique of the trope of multi-sitedness (1996, 1999: 117–35, 2004), the ability to make such connections

---

[13] As Viveiros de Castro and Goldman warn, don't try to 'understand the texts of Marilyn Strathern hastily, because they are slow, hesitant texts, folded within themselves, texts that heave and halt, and keep coming back to where they started' (2008: 25).

and to tell a story requires a perspective (scale) that is sufficiently 'big' to see all 'the parts' form; a purportedly 'global' vantage, which, revealingly, is often identical to that occupied by the 'cosmopolitan' subject (see also Castro & Goldman 2008: 36).

Whereas the goal of multi-sited fieldwork is the 'tracing [of] cultural phenomena across different settings' in order to 'reveal the contingency of what began as initial identity' (Strathern 1999: 163), then the goal of Strathern's comparisons between, say, Melanesian ownership and British patent systems (1999: 131–5), or indeed between Amazonian and Melanesian perspectivism (see Box 3.1) is altogether different. Instead of establishing putatively 'global' connections between different 'local' phenomena, as Marcus and so many other multisided fieldworkers aspire to do, Strathern seeks to foster analogies between disparate fieldwork events through what might be called a 'multi-temporal' – or, better still – 'trans-temporal' method. For the question that is implicitly raised by her postplural analytical method is what cross-cultural comparison might look like if the dimension of time itself were not to be conceived as independent from the phenomena compared; that is, if time were not assumed to constitute (as in plural metaphysics) a 'bigger scale' or transcendent position with respect to the 'things' whose comparison time facilitates.

Trans-temporal Comparison

Strathern's original fieldwork in the Mt Hagen area of the Papua New Guinea Highlands occupies a special place in her anthropological thinking. Given that the bulk of her fieldwork was carried out in the 1960s and 1970s, one might see this as posing an (automatically growing) methodological problem: does the increasingly 'historical' nature of her material not render her comparative project more and more dubious? Surely, a standard social scientific objection would go, one cannot in the same analysis compare two different ethnographic sites and two different periods. Either axis – the temporal or the spatial – must be kept stable so as to compare like with like. Strathern's response to objections of this kind (e.g. Carrier 1992a) has been characteristically subtle. Instead of trying to counter the claim that her material is not contemporary (with reference, say, to more recent stints of fieldwork), she has pleaded guilty as charged, happy to admit that many of the practices she originally observed in Hagen in the 1960ss have since changed or disappeared altogether (e.g. Strathern 1999: 142). Yet, 'the knowledge anthropologists have made out of their encounters with Melanesians ... does not cease to become an object of contemporary interest simply because practices have changed. I would indeed make it timeless in that sense' (1999: 143).

However, to simply describe Strathern's concepts as 'timeless' is perhaps not sufficiently precise a characterization of the sophisticated deployment and work of temporality in her anthropological comparisons. As we shall argue, Strathern may be said to do the same with time as with all other objects of postplural abstraction, namely to make a virtue out of its failure to act as a general scale of comparison. By treating temporality as just another thing-cum-scale of ethnographic analysis – as a scale that is no more context-independent than, say, flutes – she allows for a unique kind of comparison between societies across time. Thus her strategy is to

avoi[d] discursive connections, making a story, in order to avoid both the false negative appearance of stringing surface similarities together and the false positive appearance of having uncovered a new phenomenon. *For what the locations presented here have in common has not necessarily happened yet* ... What has not happened yet is the way in which these sites may in future connect up ... Exactly the routes that they follow, or what chains of association they set up, will be the subject of future ethnographic enquiry. (Only) the potential is present. (1999: 163; emphasis in original)

This is what we call 'trans-temporal comparison' – a distinctly postplural analytical operation at right angles, so to say, from the modernist ideal of cross-cultural comparison, and from the postmodernist penchant for multi-sited fieldwork. As we are going to show, certain writings by Strathern thus represent a concerted attempt to do an 'a-chronic' comparison across time by offering an alternative to both the synchronic project of cross-cultural comparison and the diachronic comparison of different historical moments of one society. This is why Strathern's units of comparison exist neither outside time nor are prisoners of a certain historical period. For trans-temporal comparison proceeds according to a relational logic by which the anthropologist's knowledge about certain ('Melanesian') pasts is brought to bear on certain ('Euro-American') futures by virtue of an 'abstract intensification' of particular analytically salient events during fieldwork, which Strathern calls 'ethnographic moments'.

In *Property, Substance and Effect* (1999), Strathern discusses and compares different anthropological ways of thinking 'about historical epochs as domains from which to draw resources for analysis' (1999: 145). This is why the material she gathered during her original fieldwork has not stopped being 'an object of contemporary interest simply because practices have changed' (Strathern 1999: 143). In fact, for certain analytical purposes (such as the comparative study of property rights and 'patenting') it is the other way round: the contemporary analytical purchase of her Melanesian fieldwork experiences is an effect of their non-contemporary nature: 'In certain respects "traditional" Melanesian societies belong much more comfortably to some of the visions made possible by socio-economic developments in Europe since the 1980s than they did to the worlds of the early and mid-twentieth century ... One of the times Euro-Americans may find themselves in has so to speak only just happened for them. But it may have "happened" long ago in Papua New Guinea' (1999: 146–50).

To show how this works, consider Strathern's discussion of what she calls the 'scandal' of British social anthropology's classic holistic vision, namely the at-once repulsively megalomaniac and endearingly amateurish desire to 'gather up anything' (1999: 8). For the nice thing about this holism, as she explains, is that it forces the fieldworker to be 'anticipatory... being open to what is to come later. In the meanwhile, the would-be ethnographer gathers material whose use cannot be foreseen, facts and issues collected with little knowledge about their connections. The result is a "field" of information to which it is possible to return, intellectually speaking, in order to ask questions about subsequent developments whose trajectory was not evident at the outset' (1999: 9). It is almost as if Strathern thinks that the longer the gap between fieldwork and analysis, the greater the chance that truly ethnographic insights can be reached: '[K]nowing that one cannot completely know what is going to be germane to any subsequent re-organisation of material demanded by the process of writing can have its own effect. It may create an expectation of surprise' (1999: 10).

This is what Strathern has described as 'the ethnographic moment' (1999: 3–6). Unlike its better-known sister concept of the 'ethnographic present', which has been hailed for its ability to 'transcend the historical moment' by adding more 'provisional truth[s]' to the world (Hastrup 1990: 56–7), the concept of ethnographic moment rather seems to work by *cutting away* what might, at first sight, appear to be the most likely connections between fieldwork experience and anthropological interpretation. This flies in the face of established phenomenological and largely tacit anthropological wisdom concerning the purportedly tragic loss in immediateness, sensuousness and everydayness as one's embodied memories of fieldwork experiences fade over time.

To better explain how the ethnographic moment involves a postplurally abstract process of trans-temporal scaling, we may return to our earlier visualization of postplural abstraction (see Figure 3.2).

As we explained earlier in this chapter, the logic of postplural abstraction refers to how things-cum-scales transform themselves in specific ways. As we depict in Figure 3.2, the 'ethnographic moment' may be said to constitute one such intensive self-transformation in the form of a

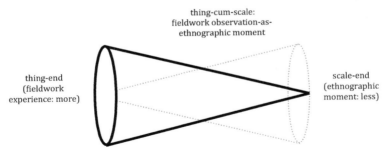

Figure 3.2 Trans-temporal comparison.

scaling of the ethnographic fieldwork observation – or, more accurately, the memorable fieldwork encounter – itself. This is the implication of Strathern's holism: the fact that the would-be ethnographer vaguely intuits that unknown future connections could one day appear, transforms her 'field of information' from being a historical artefact confined to a certain point in time (when the fieldwork took place) to a trans-temporal scale of comparison (from which analogies may be drawn at any given time). In that sense, the ethnographic moment is both more, and less, than the fieldwork event. As a postplural, intensely abstract event, it simultaneously effectuates a 'sharpening' of the anthropologist's ethnographic materials (by drawing on what is only an insignificant amount of her data), and a 'widening' of this fieldwork data by rendering visible certain 'less evident' analogies in it.[14]

14 Indeed, if trans-temporal comparison involves an act of postplural abstraction in which some 'thing' (a fieldwork experience) is 'scaled' into a different version of itself (an ethnographic moment), we may ask: Which scale is being 'thinged' in the same process? From the postplural vantage of Strathernian comparative analytics, 'time' is not different from 'flutes' in its capacity to act as a conduit for comparison: both can act as postplural scales that allow for specific kinds of relational transformations. So, on the postplural logic of trans-temporal comparison, time is reduced to just one of many (in fact, countless) possible scales for the elicitation of analogies between actual and virtual forms, and, more generally, for perception and conceptualization of the world (one could imagine an alternative universe where apples and pears are invested with the same a priori nature as time and space in Kantian epistemology). If the ethnographic moment is a certain scaling

This, then, is why 'the knowledge anthropologists have made out of their encounters with Melanesians' is "... timeless"' (1999: 143) – emphatically *not* because it is imagined to belong to a context-independent dimension of absolute and eternal truths that transcends time and space, but, very much to the contrary, because Strathern's recollections of her old Hagen fieldwork can be continuously re-activated in order to elicit ever more productive analogies with say, new forms of patent rights in the United States (1999: 132–3). After all, it could be argued that, according to non-representational anthropological theorizations of time (Munn 1986; Hodges 2008; Nielsen 2011, 2014), what happens in 'the ethnographic moment' is not restricted to 'the ethnographic present' (*sensu* Pedersen 2012c). Unlike the 'ethnographic present', which relied on a plural logic of generalized abstraction and a linear model of time, the 'ethnographic moment' allows for non-linear, 'multi-temporal' analogies to be drawn between past fieldwork experiences and present matters of comparative concern. So, if the multi-sited anthropological methods advocated by Marcus and others posit ethnographic knowledge as general but not abstract (by enabling global and plural narratives that bring together otherwise dispersed phenomena and perspectives into single stories), Strathern's comparison treats ethnographic knowledge as abstract but not general by allowing for the discovery of unlikely connections *across* times. As postplural abstractions, trans-temporal comparisons reveal links between events, which, instead of instantiating multi-sited scale shifts, work by collapsing the distinction between local and global, past and present, and other such 'plural' fictions.

It is the inherent tendency towards interpretative proliferation in ethnographic fieldwork material that makes it so important to heed the

of a 'thing-like' observation, then it is also a certain thinging of (otherwise 'scale-like') time. For more on 'trans-temporal' comparisons and analytics, see Pedersen & Nielsen (2013).

central lesson of Strathern's postplural comparative analytics: that by cutting away all the most evident relations within one's 'field of information' one becomes able to see odd but intuitive pairings of phenomena and events ('uncommon denominators'), which would otherwise be separated by space or time. For the same reason, good ethnographic descriptions require patience – the cultivation of a creative deferral, which enables the anthropologist to *not* make connections (start comparing) before the (ethnographic) moment is right. For the capacity to 'add to thoughts by narrowing them down' is not an ability with which Strathern or any other anthropologist is automatically endowed. On the contrary, putting oneself in a position to refrain from making the most obvious connections between things requires the stance of deep hesitation.

Strathern and Ontology: An Awkward Relationship

We may now return to the question that we posed at the outset of this chapter, and ask: what is the relationship between Strathern's work and the ontological turn? Clearly, profound theoretical links exist between Strathern's work and scholars associated with this approach. Due to the decisive influence Wagner has had on the development of her thinking, and the equally important influence her work exercises on that of Viveiros de Castro, Strathern is positioned at the heart of the intellectual genealogy of the ontological turn. In fact, Strathern and Viveiros de Castro for a time were involved in an intellectual exchange about the latter's theory of perspectivism, which also marked her most explicit engagement with questions of 'ontology' (see, in particular, Strathern 1999: 249–56, 2005: 138–44; Viveiros de Castro 2003, 2004; Viveiros de Castro and Goldman 2008; c.f. also Kelly 2005; Pedersen et al 2007; Vilaça 2011).

Box 3.1: Strathern's dialogue on perspectivism with Viveiros de Castro

Strathern and Viveiros de Castro have been equally explicit in recognizing the influence that they have had an each other. On his part, Viveiros de Castro recalls: 'I already read "Parts and wholes" (Strathern 1992b) before writing my article "Cosmological deixis and Amerindian perspectivism" (Viveiros de Castro 1998), but Strathernian perspectivism escaped me completely. It was only after having developed an Amazonian notion of perspectivism that I was able to make Melanesian perspectivism visible within my own conceptual aesthetic' (Viveiros de Castro & Goldman 2008/9: 29). Conversely, Strathern too has sought to detail the inspiration she has drawn from Viveiros de Castro's work. In particular, in the final chapter of *Property, Substance and Effect* (1999), she recasts her longstanding interest in Melanesian kinship in the language of ontological differences between the perspectives different kinship positions afford. As she points out in this connection, 'the evocation of ontology is quite deliberate here. For what lies behind this description is Viveiros de Castro's concern with the primitive ontological base [on which...] much anthropological exegesis rests' (1999: 251). In other words: the only thing that may be deemed 'primitive' in anthropological parlance is the impoverished framework upon which anthropological analyses sometimes rest; and Viveiros de Castro's invocation of ontology offers one possible means of rectifying this. As Strathern writes, what 'attracted me to Viveiros de Castro's Amazonian perspectivism is the clarity with which he locates it as a matter of ontology not epistemology. It is not about what one knows but about how one is, about the nature of the body with which one inhabits the world and apprehends it' (2005: 140).

To be sure, true to her persistent focus on ethnographic specificity, Strathern has stressed that Viveiros de Castro's theory of perspectivism cannot be applied directly to Melanesian ethnographic contexts where the 'division between humans and others is not the principal perspectival axis' (1999: 232; see also 2005: 140), as it is in the Amerindian ethnographic contexts discussed by Viveiros de Castro (more on which in the next chapter). Nevertheless, on several occasions she has also emphasized how Viveiros de Castro's model of perspectivism has offered her 'a useful re-entry into the Melanesia material' (1999: 252; see also 2005: 138–44). After all, as she puts it, 'there are *ontological consequences* to being a son to these people and a sister's son to those, or to being a consanguine by contrast with an affine' (1999: 252–3; emphasis added). More precisely, at issue here is the ethnographic as well as analytical question of what perspective 'might ... look like in a society that does *not* ... imagine perspectives as self-referential, unique "contexts" for action and hence with the potential to co-exist with, and overlap with, limitless numbers of "unique" others' (1999: 249). It is precisely this shared difference from 'Euro-American' assumptions about what perspectives might be that in Strathern's view delineates a fruitful (postplural) scale of comparison between her own 'Melanesians' and Viveiros de Castro's 'Amerindians'. In particular, Melanesia and Amazonia provide conceptual vantage points that appear equidistant from the plural notion of what Strathern calls 'perspectivalism', according to which determinate things, understood as objects of attention, can be seen and compared from different points of view that therefore also provide varying 'contexts' for them. Contrasting this characteristically 'Euro-American' notion of perspective with her own perspectivist analysis of Melanesian kinship, she queries: 'what would be finite here? Could it be the manner in which one's perspective was returned to one?' (1999: 249). As she stresses, these, among other decidedly ontological questions about what might *count* as a perspective in any given ethnographic situation, are what Viveiros de Castro's work on Amerindian perspectivism allowed her and others to ask.

At the same time, however, one is also left with the unmistakable impression – based as much on what she has *not* said than what she has said about ontology and key scholars associated with it, such as Viveiros de Castro – that Strathern does not in any full sense consider herself as part of this anthropological 'turn'. Perhaps, to borrow her own way of capturing the productive tension between anthropology and feminism, the relationship between Strathern and ontology can be characterized as irreducibly 'awkward'? But where does this awkwardness lie? To be sure, it can have absolutely nothing do to with Strathern's core notion that 'relations' are not to be found out there, but are rather something that 'we' cannot help doing as anthropologists using the English language when conducting our analyses of 'them'. Nor, and no less emphatically, has it to do with the fact that, as a logical consequence of this relational analytics, everything is imagined *as if* it were a comparison – the postplural injunction, that is, that any given thing contains the latent potential to be 'scaled' into what it is not (yet). And, finally, it cannot have anything to do with the fact that Strathern's interests are explicitly epistemological insofar as they amount to a systematic investigation and critique of the most basic conditions of possibility of what anthropological knowledge is, including potential 'persuasive fictions' (Strathern 1987b) through which new and more adequate ethnographic descriptions might be made. After all, as we explained in the Introduction, each of these three basic premises for conducting anthropology (that relationality is an anthropological invention rather than an ethnographic fact, that comparative activities have ontological ramifications, and that epistemological questions in anthropology are inseparable from ontological ones) lies at the core of the ontological turn as we understand it.

In sum, within the premises of Strathernian anthropology and 'our' ontological turn more generally, the business of anthropology is not comparison of ontologies (c.f. Descola 2013) but rather comparison *as* ontology (c.f. Holbraad et al. 2014). This goes to the core of her concept of the relation as a form of anthropological enquiry, that by comparing something to something else one gets to the heart of them both. And whatever one thinks about that as a vision for anthropology, the stakes it raises are unmistakably ontological in the sense this term is deployed in this book, pertaining to the question of what 'things' – that is, objects of anthropological attention – *could be.* To reiterate an observation already made in the Introduction, this is just another way of saying that the ontological turn is at heart an methodological project, which strives to take seriously the ontological effects of engaging in ethnographic practice and anthropological knowledge- (and concept-) making – thus merging epistemology with ontology, as we saw.

This is not to say that Strathern's anthropology and the kind of ontological turn that concerns us in this book are identical projects. Several differences, pertaining to choice of terminology and the manner (including the directness) of writing, may be detected between the two analytics. Indeed, we suggest in closing, what makes the relationship between Strathern and the ontological turn awkward is the deliberately direct way in which certain proponents of this approach have set up questions of ontology as anthropological challenge. After all, as we have sought to show in this chapter, Strathern's overarching anthropological project may be boiled down to an inherently feminist vision for a new aesthetics of ethnographic description (or 'redescription', as Ashley Lebner has recently coined it, 2016) that is deliberately non-explicit, and systematically non-transparent, about its own theoretical ground. Small wonder that Strathern has not wholeheartedly embraced 'ontology', for doing so might be seen to constitute a breach with the principled way in which she has always managed to defer completeness in her work. Being too transparent about the theoretical, metaphysical and philosophical ramifications of anthropology, might, from the perspective of her and other feminists, such as Haraway, put hesitation, and therefore surprise, at risk as analytical, political and ethical endpoints.

At first glance, it seems to be very much the other way around for someone like Viveiros de Castro and other scholars associated with the ontologial turn, ourselves included (c.f. Lebner 2016). After all, to him and likeminded scholars, ontological questions and their philosophical implications cannot be emphasized enough, for it is precisely in this assiduous and relentless explication of its inevitable metaphysical dimensions that anthropology's unique theoretical contribution as well as political potential lies. Perhaps Strathern has been hesitant towards some of the language that has been used to promote an ontological turn partly because its declarative style and assertive tone seems to be reaching for the kind of analytical completeness Strathern's feminism deliberately avoids. Thus understood, Strathern is neither fully part of the ontological turn nor fully outside it. Which, indeed, is a quite Strathernian place to find oneself in. After all, as we discussed in Chapter 1, several of the scholars associated with different versions of the ontological turn have quite deliberately presented it as an attempt systematically to delineate alternative ontological regimes (e.g. Descola's fourfold scheme, Evens's nondualism), and even proponents of the strictly methodological version we seek to articulate in this book have also on occasion flirted with such total-sounding ideas as 'multiple worlds', 'ontological pluralism', political ontology, and so on. Indeed, as we shall see in the next chapter, in articulating most explicitly in ontological terms the reflexive line of thinking we consider to lie at the heart of the ontological turn, Viveiros de Castro himself has sometimes presented it in such 'totalizing' and assertive language.[15] Clearing up the confusions that this ambiguity has produced is part of our motivation in providing our own exposition of Viveiros de Castro's debt to Strathern's extreme analytical reflexivity in the chapter that follows. For, as we are going to show, Strathern's analytical method and her concept of the relation in particular is not just constitutive for but entirely congruent with Viveiros de Castro's attempt to introduce the question of ontology as an anthropological concern.

---

[15] There is no reason why it should not be possible to imagine an ontological turn that systematically deploys hesitation and incompleteness and other 'feminist' (in the particular sense of the term discussed in this chapter) descriptive and analytical aesthetics (for a potential candidate, see Tsing 2014).

FOUR

Natural Relativism: Viveiros de Castro's Perspectivism and Multinaturalism

'The West is dead. Get over it.' So goes a recent Facebook update by Eduardo Viveiros de Castro, Brazilian anthropologist, political activist, and father of anthropology's ontological turn. More than anyone else associated with this 'turn', Viveiros de Castro has been the target of admiration and praise, but also relentless critique and ingrained scepticism, for tabling ontology as a matter of anthropological concern. Small wonder, considering his provocative style. In its programmatic content and citation-friendly form, the preceding comment is very much characteristic of Viveiros de Castro's work – pithy, and posted on social media.[1] Indeed, Viveiros de Castro's reputation as something of a maverick in anthropological and other academic circles is owed not just to the style of his interventions, but also to the way in which he has promoted his argument about ontology in explicitly political terms. Anthropologists, for him, are above all conceptual freedom fighters, as he explains in a much cited (and debated) passage:

For many of us [who became adults around 1968] anthropology was and still is an insurrectionary, subversive science; which fought for the conceptual self-determination of all the planet's minorities, a fight we saw as an indispensable accompaniment to their political self-determination.

---

[1] Viveiros de Castro has a strong web presence, with tens of thousands of followers on Twitter and Facebook.

Anthropology is consistently guided by this one cardinal value: work-ing to create the conditions for the conceptual, I mean ontological, self-determination of ... the world's peoples, and ... it is thus a political science in the fullest sense. (2003: 16)

Perhaps, much as we suggested regarding Marilyn Strathern and the 'crisis of representation' in the last chapter, Viveiros de Castro's project can be glossed as what postcolonial anthropology would have looked like had he been in charge of its management. On the one hand, Viveiros de Castro thus shares with, say, Edward Said (1978) the goal of provincializing the West through a critique of its often tacit hegemonies of representation and other forms of ideological domination. Still, on the other hand, his anthropological-cum-political project also differs from this and other visions for producing 'theory from the south' (Comaroff & Comaroff 2011) in crucial ways, including the fact that he writes about seemingly non-postcolonial ('orientalist', 'Victorian') topics as culture, cosmology and, most jarringly perhaps from a subaltern perspective, ontology.

And yet, for Viveiros de Castro this is very much the point. His argu-ment is that returning to just these topics, and particularly addressing the ontological (as opposed to merely the epistemological) stakes that underwrite them, is necessary in order to develop for anthropology the kind of emancipatory political agenda the critical literature on postco-lonialism seeks to promote. It is a manner, as he often puts it, of 'taking seriously' the people that we study as anthropologists, without, that is to say, encasing the challenges that they pose to us in our own ideas about 'beliefs', 'worldviews', 'cultures', 'epistemes', and the like. As Latour has put it, Viveiros de Castro's vision is thus to use Amerindian cosmologies and the practices they inform 'as a bomb with the potential to explode the whole implicit philosophy so dominant in most ethnographers' inter-pretations of their material' (2009: 2). Crucially, if such it is, this 'bomb' is not thrown just to further the interests of anthropology or philoso-phy, but also, and for Viveiros de Castro most potently, for the political purpose of furthering the interests of Amerindian and other indigenous peoples. Ontology, for him, is what anthropology has always needed in order to fulfil its untapped potential for radical conceptual politics, to 'kill the West', for good.

In the previous two chapters we showed how Wagner and Strathern prepared the theoretical and methodological ground for what we consider to be the defining analytical move of the ontological turn, namely the injunction to treat ethnographic contingencies reflexively and experimentally, as the source of anthropological conceptualization. Still, although the turn towards ontology can thus be traced back to certain theoretical developments within the Boasian tradition of American cultural anthropology (epitomized in the work of Wagner) as well as structural functionalism in the British tradition of social anthropology (to which Strathern is an heir), it was Viveiros de Castro who first introduced 'ontology' as a matter of concern for contemporary anthropology.[2] If Wagner set the ontological turn on its tracks by reinventing the American tradition of symbolic anthropology and Strathern radicalized it by transforming Radcliffe-Brown's focus on 'social relations' into a new language for ethnographic description, then, as we demonstrate in this chapter, Viveiros de Castro makes explicit the ontological stakes of this way of thinking by refiguring it in terms of the French structuralist and poststructuralist concern with transformation and self-differentiation.

So it is with Viveiros de Castro that the ontological turn truly comes into its own. In what follows, we explain how this came about by charting the emergence of ontology as a concern in his work on Amerindian animism in the 1990s and 2000s. As an explicit methodological agenda for anthropology, we show, the concern with ontology developed out of the analysis of Amerindian ethnography since, as Viveiros de Castro puts it,

---

[2] Of course, as pointed out in previous chapters, the term 'ontology' has been invoked by numerous anthropologists before Viveiros de Castro, and perhaps most notably by Hallowell (1960), to different purposes.

the argument about ontology renders explicit the 'metatheoretical premises' of his earlier ethnological analysis (2013: 473). Accordingly, we first present the rudiments of Viveiros de Castro's now famous analysis of Amerindian 'perspectivism', showing how it effectively 'intensifies' the relational premises of Lévi-Straussian structuralism by bringing to bear on them a Strathernian conception of the relation. In this connection we address also the influence of the philosophy of Gilles Deleuze on Viveiros de Castro's analysis of Amerindian cosmology and explore how these arguments play themselves out in relation to the key concept of 'multinaturalism', and the method of 'controlled equivocation' he builds upon it. One of our main aims in this discussion will be to show why Viveiros de Castro's approach amounts to *anything but* the kind of exoticizing essentialism with which his critics so often charge him.

Amerindian Perspectivism

The theory of perspectivism, it is widely accepted (e.g. Course 2010; Kohn 2015),[3] represents Viveiros de Castro's most important and influential contribution to anthropology. To get a proper handle on it, and thus avoid repeating superficial accounts available in the critical literature (e.g. Ramos 2012), we may begin by noting that perspectivism plays a role in Viveiros de Castro's thinking which is not dissimilar to that of ceremonial gift exchange in Marilyn Strathern's work. For, much like her concept of 'Melanesian dividuals' (1988) is both grounded in her own fieldwork among a specific people in the Papua New Guinean Highlands as well as in a synthesis of a range of other anthropological work from elsewhere in Melanesia, so Viveiros de Castro's model of

---

[3] In the words of Magnus Course, perspectivism 'has now become the dominant paradigm (some might even say orthodoxy) within which most Brazilian, European, and an increasing number of North American anthropologists concerned with the region are working ... Its influence ... lies in its ability to make sense of a wide variety of ethnographic facts' (2010: 249).

Amerindian perspectivism is derived from his own work among the Araweté people, in the Xingu River region of Brazilian Amazonia (1992), as well as an extensive reading of other ethnographic scholarship from the Amazonian region at large, and not least of the work of graduate students who worked with him throughout the development of the model (e.g. Carneiro da Cunha & Viveiros de Castro 1993; Viveiros de Castro & Fausto 1993; Viveiros de Castro 1995). In fact, and again much as with Strathern's discussions on gender and gifts (1980; 1988), Viveiros de Castro's theory of perspectivism brings together two specialized and partly disconnected literatures, namely, on the one hand, the extensive ethnographic record on Amerindian cosmologies, mythologies and rituals, and, on the other, the no less sizable work on 'animist' cosmologies among hunter-gatherers and other so-called 'indigenous' peoples around the world. Indeed, Viveiros de Castro's first impact on international anthropology was as a central contributor in what eventually became known as the 'new animism' (Harvey 2005), in debate with his fellow Amazonianist Philippe Descola (1992, 1996, 2013) and circumpolar North scholar Tim Ingold (1986, 1998, 2000), among others (Howell 1989; Bird-David 1999; Pedersen 2001; Willerslev 2004). We may begin our account of perspectivism, then, by showing how it emerged as a response to the debate about animism.

Viveiros de Castro takes his original lead from Descola, for whom animism is a 'system' of human thinking that can be found to various degrees and in different mixtures in all societies. In fact, 'animic systems are ... a symmetrical inversion of totemic classifications: In totemic systems non-humans are treated as signs, in animic systems they are treated as the term of a relation' (Descola 1996: 87–8). Thus, Descola goes on to argue, animist cosmologies are premised on a 'continuity between humans and non-humans' (1996: 89). In a similar vein (and in direct engagement with Descola, Durkheim and Mauss, as well as Lévi-Strauss), Viveiros de Castro defines animism as 'an ontology which postulates the social character of relations between humans

and nonhumans: the space between nature and society is itself social'
(1998: 473). But this is not all. Elaborating on a pan-Amerindian scale
ideas first developed in his research on sixteenth century Tupian can-
nibalism as well as Araweté war songs (1992; see also Stolze Lima
1999), Viveiros de Castro charts the subtlety with which this 'continuity' between humans and other species is expressed in Amerindian
cosmologies and the practices that they inform (including myth, shamanism, hallucinations and dreaming, hunting, the focal role of affines
in kinship, and other fields in which the nature of social relationships
is prominently articulated). In particular he seeks to demonstrate
that, more than just positing a social continuity between humans and
non-humans where naturalism would posit an ontological divide,
Amerindian animism requires a thoroughgoing, and thoroughly intricate, analytical reversal of the coordinates of the 'Western naturalist
ontology' with which anthropologists habitually operate, fundamentally reshuffling and destabilizing core binaries such as the social/
cultural versus the natural, the material versus the mental/spiritual,
and indeed the epistemological versus the ontological.

It is for this 'stronger' (Pedersen 2001) form of animism that Viveiros
de Castro reserves the word perspectivism. Having elaborated his analysis with reference to an overview of Amazonianist ethnographic literature in a series of works in the late 1990s, in a later article he summarizes
the salient features of perspectivism as follows:

['Perspectivism' is] a label for a set of ideas and practices found throughout indigenous America and to which I shall refer, for simplicity's sake, as
though it were a cosmology. This cosmology imagines a universe peopled
by different types of subjective agencies, human as well as nonhuman, each
endowed with the same generic type of soul ... which determines that all
subjects see things in the same way. In particular, individuals of the same
species see each other (and each other only) as humans see themselves, that
is, as beings endowed with human figure and habits, seeing their bodily
and behavioural aspects in the form of human culture. What changes when

passing from one species of subject to another is ... the referent of these concepts: what jaguars see as 'manioc beer' (the proper drink of people, jaguar-
type or otherwise), humans see as 'blood'. Where we see a muddy salt-lick
on a river bank, tapirs see their big ceremonial house, and so on. Such difference of perspective – not a plurality of views of a single world, but a single
view of different worlds – cannot derive from the soul, since the latter is
the common original ground of being. Rather, such difference is located in
the bodily differences between species, for the body and its affections (in
Spinoza's sense, the body's capacities to affect and be affected by other bodies) is the site and instrument of ontological differentiation and referential
disjunction. (2004: 6)

Situated at right angles to the cosmological coordinates of Latour's 'modern
constitution' (1993), Amerindian perspectivism shifts the conventional
naturalist concepts of multiculturalism and (mono)naturalism,
turning them into what Viveiros de Castro brands 'monoculturalism'
and 'multinaturalism' (1998: 478). While multiculturalism, which in
anthropology has typically taken the form of cultural relativism and what
he dismisses as 'constructivist epistemologies' (as we also discussed in
Chapter 1 in relation to parallel developments within the field of Science
and Technology Studies, STS), is premised on the 'unity of nature and
the plurality of cultures', Amerindian perspectivism is based on 'spiritual
unity and corporal diversity' (1998: 470) – one culture, many natures
(see 2004: 6). Perspectivism, in other words, is exactly the opposite of
relativism as we know it – the notion that 'perspectives' (or 'cultures', or
'worldviews') are subjectively relative in the sense that they differ from one
human subject (collective or individual) to another. Instead, the relativity
in question here is one of objective positions, which vary according to the
different ways in which a (universally available) point of view is embodied.
As Viveiros de Castro spells it out, 'cultural relativism imagines a diversity
of subjective and partial representations (cultures) referring to an objective
and universal nature, exterior to representation. Amerindians, on the other
hand, propose a representative or phenomenological unity that is purely

pronominal in kind applied to a real radical diversity' (2004: 6). Hence the oft-cited passage:

[H]umans see humans as humans, animals as animals and spirits (if they see them) as spirits; however animals (predators) and spirits see humans as animals (as prey) to the same extent that animals (as prey) see humans as spirits or as animals (predators). By the same token, animals and spirits see themselves as humans: they perceive themselves as (or become) anthropomorphic beings when they are in their own houses or villages and they experience their own habits and characteristics in the form of culture. (1998: 470)

At the heart of the matter, in other words, is a sort of *natural relativism* (2013: 498), as the title of this chapter would have it – 'multinaturalism', that is, as an ontological matter of concern: The jaguar sees the world in the same way *as* humans. *What* the jaguar sees, however, differs from what the hunter sees.

This destabilization of the naturalist distinction between nature and culture also warps the distinction between body and soul, as shown by the extensive ethnographic materials from the Amazonian region that Viveiros de Castro reviews. 'In Amerindian cosmology', he writes, 'there are no points of view *onto* things, things and beings are the points of view *themselves*' (2004: 11). It follows that to see a different world it takes a different body, understood, crucially, not as a physical or material object, but as the dynamic assemblage of 'affects, dispositions or capacities which render the body of every different species unique' (1998: 478). It is because of their different bodies, and not least their differing dispositions towards each other as either predators or prey, that jaguars in perspectivist cosmology see beer where humans see blood, as in the afore-cited example. Indeed, this is also why the possibility of travelling from one species perspective to another, which is what shamans are so crucially able to do, depends on bodily, rather than spiritual, transformation. This is so because, within Amerindian animist cosmology, bodies are a form of 'clothing', and shamans don masks and other bodily garbs in order to be able to transform themselves into spirits and/or animals:

The animal clothes that shamans use to travel the cosmos are not fantasies but instruments: they are akin to diving equipment, or space suits, and not to carnival masks. The intention when donning a wet suit is to be able to function like a fish, to breathe underwater, not to conceal oneself under a strange covering. In the same way, the 'clothing' which, amongst animals, covers an internal 'essence' of a human type, is not a mere disguise but their distinctive equipment, endowed with the affects and capacities which define each animal. (1998: 482)

The body, in short, is the site of perspectival differentiation, rather than the object of different perspectives (as it is for cultural relativism or social constructivism). Conversely, the animist concept of soul is the bearer of the one attribute all these mutually differing species have in common, namely the capacity to embody a perspective in the first place. According to this conception, then, 'the soul' is an intrinsically relational and relative category, or, to adopt the more technical vocabulary favoured by Viveiros de Castro in his early writings, it is *deictic*:

Whatever possesses a soul is a subject, and whatever has a soul is capable of having a point of view. Amerindian souls, be they human or animal, are thus indexical categories.... Such deictic 'attributes' are immanent to the viewpoint, and move with it ... Salmon are to (see) salmon as humans are to (see) humans, namely (as) human. (1998: 469)

By borrowing this terminology from linguistic theory, Viveiros de Castro seeks to articulate how, in Amerindian animism, the soul is not conceived of as a species-specific essence (whether material or immaterial) but as a formal attribute pertaining to otherwise varied relationships between humans and non-humans. The soul is the 'I' that points indexically to all the different bodies that can potentially occupy the position of 'you' or 'it'.[4] It is precisely because of this ontological – or 'hyper' (Course

---

[4] Magnus Course draws out the consequences with respect to linguistic theory: 'In a conventional use of the term, deixis refers to the referential meaning of an utterance being dependent on the spatial, temporal, or personal position from

2010) – deixis that the animist cosmologies found in Amazonia can be described as monoculturalist. For the 'spiritual unity' across which the traffic in perspectives moves is nothing other than the indexical position of the subject (the 'I'), that is, the first person deictic attribute, which on the analogical logic of animism is extended to include all human and non-human persons perceived to be embodying a point of view. In sum, 'the spirit or soul (here not an immaterial substance but rather a reflexive form) integrates, while the body (not a material organism but a system of active affects) differentiates' (Viveiros de Castro 1998: 479). So, the soul is the vehicle of an ontological deixis involving the perception of and traffic between multiple natures, each of them tied to distinct but dynamic bodily affects and capacities.

Savage Structuralism

But is multinaturalism not simply a new version of Lévi-Straussian structuralism? Certainly this is one the criticisms that is often levelled at Viveiros de Castro's model of Amerindian perspectivism, namely that it reproduces the very binary distinctions it is supposed to be repudiating, and not least the arch-structuralist (or even arch-modern, 'Western') distinction between nature and culture. Fuentes and Kohn put the point judiciously: 'the multiplication of natures is not an antidote to the problem posed by the multiplication of cultures. For this only sidesteps the hard question: can anthropology make general claims about the way the world is?' (2012: 139). In his own forceful and much-cited critique, Terrence Turner pushed further the idea that Viveiros de Castro's model reproduces the binaries of structuralism since it 'turn[s] Lévi-Strauss's [reduction of culture to nature] inside out through an equally

---

which it is emitted. Yet in the deixis characteristic of perspectivism it is the world itself which is dependent on the position from which its perception emanates' (2010: 25).

radical but opposite reduction of nature to culture, achieved through the elevation of subjective perspective over objective associationism as the determining constituent of the "spiritual" identities of all creatures, animals and humans alike' (2009: 11). At any rate, as Course puts it, the worry persists among some commentators that 'despite its attempt to free itself from such intellectual baggage in the description of a radically different Amerindian ontology, [perspectivism] remains at least partially imprisoned within this infernal dichotomy and its accompanying ontological assumptions' (2010: 253).

Without wanting to enter into a region-specific discussion about how accurately the theory of perspectivism conveys Amerindian and other ethnographic realities,[5] it seems to us that, damaging as some

---

[5] A large and continuously growing body of scholarship documents the presence of perspectivist ideas around the world. In so-called indigenous America alone, this includes (in addition to the substantial literature mentioned in Viveiros de Castro's own writings on this topic, 1998 and 2012), Vilaça (2005, 2015), Kohn (2008, 2013), de la Cadena (2010), Fortis (2010), Pitarch (2011), Bonelli (2012), and Uzendosky (2012), among others. On the purchase of perspectivism as an ethnographic as well as a theoretical category in Inner Asian and North Asian contexts, see Pedersen, Empson and Humphrey (2007), as well as Brightman, Grotti and Ulturgasheva (2014), Charlier (2015), Pedersen (2001, 2011), Willerslev (2004, 2007), Stépanoff (2009); Pedersen & Willerslev (2012); Swancutt (2012); and Willerslev and Pedersen (2010). Other recent anthropological studies discussing perspectivist ethnography and theory include Stasch (2009), Holbraad (2012), Bubandt (2014) and Candea (2012). In view of the enormous ethnographic variation these studies represent, including the fact that each is the product of complex historical and political processes, one may question whether the invocation of the perspectivist model is equally productive in each case. It certainly would be a fundamental misunderstanding of the ontological turn to assume that perspectivism can take only one form. On the contrary, the purchase of this concept within a comparative anthropological framework depends on its capacity to be 'stretched' or 'scaled' through its encounter with new ethnographic contingencies and anthropological matters of concern (see also Chapter 3). Take, as an example of this conceptual 'stretching', the Inner Asian case. Here, several anthropologists (Pedersen 2007; da Col 2007; Humphrey 2007; Kristensen 2007; Broz 2007; Charlier 2012; cf. Humphrey 1996; Pedersen 2001) have shown how perspectivism involves shamans' capacity both to transgress human/non-human divides (as in the Amazon) and embody multiple kinship and gender positions across an interhuman ontological divide (as in Melanesia, cf. Strathern 1998: 249–60; see also Box. 3.1). Unlike both the Melanesian and the Amazonian cases, however, social relations in Inner

of these criticisms might appear at first sight, they tend to overlook a crucial, even defining, aspect of Viveiros de Castro's theoretical model. We are referring to the fact that, for him and other scholars working on perspectivism in Amazonia (e.g. Villaça 2005; Kelly 2011), the nature/culture binary is not so much reversed or overcome (*sensu* Descola 2014), but rather 'distorted' to certain theoretical ends. For even if Viveiros de Castro's ultimate goal is 'to produce a non-dualist conceptual alternative to the conceptual dualisms that organize and constitute the field of anthropology: individual and society, nature and culture, traditional and modern' (Viveiros de Castro & Goldman 2009: 32), he is all too aware that the only way to do so, paradoxically, is by experimentally reconceptualizing these and other binaries, to bring them into contact with their own 'limit' (Viveiros de Castro 2013: 481–3) – not, crucially, to 'go beyond' or 'transcend' the binaries, but in order to get still closer to that limit, cultivating ways of conducting one's thinking *within* its conceptual threshold (Viveiros de Castro 2014: 86; cf. Agamben 2004). And the reason for this, as we shall now see, is that for Viveiros de Castro the 'binary oppositions' of structuralism are better theorized as Strathernian relations.

Unlike many anthropologists of his generation, Viveiros de Castro is not shy about his structuralist heritage. On the contrary, he flags his Lévi-Straussian credentials and 'Parisian' inspirations at every opportunity. In fact, surprising as this may seem to younger generations of scholars, particularly in the English-speaking tradition, reared in the dogma

Asia tend to be irreducibly hierarchical, no matter whether we are talking about relations between humans, between humans and nonhumans, or between different nonhumans. So, whereas in Amazonia, 'exchanges of perspectives between different kinds of beings are conceived in thoroughly "horizontal", or symmetrical, terms', in Inner Asia 'changes of perspective frequently take place in ways that are best described as "vertical"' (Holbraad & Willerslev 2007: 330). It is because of this core ethnographic contrast between horizontal and vertical cosmologies, along with other historically generated cultural differences, that Viveiros de Castro's model has to be significantly modified if not wholesale transformed when put to work in Inner Asia (ibid.; Pedersen 2011).

that 'French structuralism' is not just epistemologically and methodologically but also politically and ethically flawed, it was, more than anything, Lévi-Straus's *Mythologiques* that provided Viveiros de Castro with
his most decisive ethnographic and theoretical inspiration. In post-1968
Rio, he writes, 'the expression "*la pensée sauvage*" did not signify "the
savage mind." To us it meant untamed thought, unsubdued thought,
wild thought. Thought against the State, if you will. (In remembrance of
Pierre Clastres)' (2003: 3).

Still, Viveiros de Castro's anthropological project differs from that of
Lévi-Strauss in important respects, notably his treatment of the problem of human nature. For if Lévi-Strauss, especially in the early years
of the development of structuralism, grounded his theory of structural
transformation in a purported 'psychic unity of mankind', then Viveiros
de Castro takes this model in the opposite direction by decoupling – or
could we say liberating – it from any fixed natural ground. The moment
one reads Lévi-Strauss in this way,

'human nature' – ... the darling concept of the third great anthropological
tradition [i.e. the French] – would ... stop being ... a self-same substance
situated within some naturally privileged place (such as the brain, for example). Instead, nature itself would be accorded the status of differential relation ... If culture is a system of differences, as the structuralists liked to say,
then so is nature: differences of differences. (Viveiros de Castro 2013: 482)

So, as Viveiros de Castro has himself discussed (2010, 2014), structuralism
and poststructuralism are continuous with each other in this respect:
poststructuralism is an avatar, precisely, of the modalities of self-
transformation structuralism implies, which came to the fore particularly
in Lévi-Strauss's mature studies of myth in the *Mythologiques*.

Indeed myth is a central preoccupation also in Viveiros de Castro's
own account of Amerindian perspectivism. It is important above all, for
him, because it presents the conditions of possibility of the perspectivist
parsing of the continuous and the discontinuous by effectively collapsing

the two into each other. As he had argued already in his original exposition of Amerindian perspectivism, myth is the 'vanishing point' where the differences between points of view on which perspectivist cosmology depends are at the same time annulled and exacerbated:

In myth, every species of being appears to others as it appears to itself (as human), while acting as if already showing its distinctive and definitive nature (as animal, or plant or spirit). ... [This is] a state of being where bodies and names, souls and affects, the I and the Other interpenetrate, submerged in the same pre-subjective and pre-objective milieu – a milieu whose end is precisely what the mythology sets out to tell. (1998: 483–4)

In more recent discussions of perspectivism (2007, 2010, 2014), a central preoccupation of Viveiros de Castro has been to avoid the familiar anthropological notion that the origin myths of indigenous cosmologies posit an 'original state of indifferentiation' – a continuous magma out of which the differentiated elements of the cosmos as we know it emerge, which is often depicted, for example, in Polynesian mythology (cf. Schrempp 1992). To imagine the Amerindian pre-cosmos as such a state of indistinction would weigh the scales on the side of spiritual continuity at the expense of the embodied discontinuities that perspectivism enunciates, rendering the differences between species logically as well as chronologically derivative – a logically discontinuous passage, if you like, from the continuous to the discontinuous. To the contrary, argues Viveiros de Castro, Amerindian myths abidingly emphasize the *continuity of differentiation* – an insight he finds also at the heart of Lévi-Strauss's own emphasis on transformation in the *Mythologiques* (Viveiros de Castro 2010; 2014):

[T]he actants of origin myths are defined by their intrinsic capacity to be something else; in this sense, each mythic being differs infinitely from itself, given that it is posited by mythic discourse only to be substituted, that is, transformed. It is this self-difference which defines a 'spirit', and which makes all mythic beings into 'spirits' too... In sum, myth posits an ontological regime commanded by a fluent intensive difference which incides on each point of a heterogenic continuum, where transformation is anterior to form, relation is superior to terms, and interval is interior to being. (2007: 158)

Amerindian myths, then, enunciate the ontological conditions of possibility both of the discontinuities between species that are characteristic of the post-originary world in which perspectivism operates *and* of the fact that, by virtue of the spirit-form of their souls, the differences between these species are nevertheless continuous with each other, i.e. of the same type, since they are all differences of (in fact, as we saw, *as*) perspective. This logical merger of the continuous with the discontinuous puts at the heart of perspectivism and animism the possibility of the trans-species metamorphoses so often depicted in myth and enacted in shamanism, dreaming, hunting and so forth. Thus ontological perspectivism rests on the 'mythical' principle that each being in the cosmos has the potential to transform into every other because all beings contain each other's perspectives immanently. Beings can 'become-other' because they already 'are other:' they are constituted as beings by their very potential to become something else (see also Holbraad & Willerslev 2007).

It is this concept of self-differentiation – the lynchpin of his theory of perspectivism – that represents the point of closest contact between Viveiros de Castro's model of perspectivism and Strathern's concept of the relation. Viveiros de Castro's structuralism is effectively forti-fied by the Strathernian idea that relations are not to be understood as just objects of investigation (e.g. as the building-blocks of social or semiotic structures) but always also as the *means* of inquiry.[6] As we saw

---

[6] As Viveiros de Castro puts it himself, 'Strathernian anthropology is the most sophisticated theory of the relation that our discipline has produced since Lévi-Strauss's structuralism' (Viveiros de Castro & Goldman 2008/9: 24). More precisely, for Viveiros de Castro, it is Strathern's refusal to render relations substantial and observable as data 'out there' that makes her concept of the relation so theoretically

in the previous chapter, Strathernian relations operate as descriptive-cum-analytical 'postplural scalings' through which the varying forms and contents that social, semiotic, and other observable 'relations' can take in any particular ethnographic instance produce a reciprocal effect on the anthropologist's analytical understanding of what a relation might be in the first place. Relations are in others words conduits, as we explained, through which 'native' forms of social life get transformed into anthropological manners of conceptualizing relations.

It follows that if Amerindian cosmology comprises entities whose defining characteristic is their capacity to differ from themselves, then we may conclude not just that these entities are, actually, relations, but also that they are relations of a particular kind, namely relations that have the capacity to differ from themselves, *also qua relations*. They are relations, then, whose character is peculiarly open-ended, since their definition – the ontological question of *what they are* – remains constitutively unspecified, always liable to change. A bit like with Heidegger's notion that Dasein is the form of being for which being is an issue (1996), self-differentiation implies relations for which relation is an issue, since 'what varies ... is not the content of relations, but rather the very idea of relation' (Viveiros de Castro 2013: 481). Self-differentiation implies, in other words, Strathernian relations – that is, to recall a formulation from the previous chapter, relations not as what exists between people 'out there', but as what anthropologists are compelled to invent in order to study the people in question.

So this is why, in Viveiros de Castro's treatment, the distinction between nature and culture does not feature as an opposition between two extensive categories, or essences, but rather provides the coordinates that mark out a particular set of potential transformations – a structure,

subtle: 'anthropology distinguishes itself from other discourses on human sociality, not by holding any firm doctrine about the nature of social relations, but on the contrary, by maintaining only a vague initial idea of what a relation might be' (Viveiros de Castro 2013: 483).

yes, but only if understood in a thoroughly (late) Lévi-Straussian fashion, as a matrix of transformations (Viveiros de Castro 2014). Nature
versus culture, in other words, is a Strathernian relation insofar as it is
treated as containing within itself – within its threshold – the horizons
of its own transformation into something else. Inasmuch as it shifts us
from ordinary mononaturalist/multiculturalist assumptions towards
novel ontological possibilities such as the multinaturalist/monocultural-
ist perspectivist cosmos, Viveiros de Castro's argument proves his point
about the self-differential character of the 'binary relations' (as opposed
to binary oppositions) of structuralism. And as we are now about to see,
it also has a transformative effect on the very way we conceive of anthropological inquiry.

Anthropology as Ontology

Speaking to the core agenda of this book, one could put matters very
simply: the ontological turn in anthropology came about with Viveiros
de Castro transposing systematically onto the conduct of anthropological
inquiry the ontological questions that his analysis of Amerindian
animism posed. Indeed, just as Wagner presented *The Invention of
Culture* as the anthropological 'epistemology' that corresponded to the
ethnographically driven argument of his earlier book *Habu*, as we saw in
Chapter 2, Viveiros de Castro presents the main tenets of the ontological
turn in his landmark article 'The relative native' (2013 [2002]), as a
manner of rendering explicit what he calls the 'metatheoretical premises'
(2013: 473) of his earlier arguments about Amerindian perspectivism.
In this way, Viveiros de Castro's 'Amazonianization' of anthropology
consists in extending into the heart of anthropological thinking and
theorizing the core tenet of Amerindian perspectivism, namely, the idea
that differences between 'perspectives' are to be seen in ontological rather
than 'merely' epistemological terms. This 'merely' is important. As we
have sought to make clear from the outset of the book, the point of the

ontological turn is not to opt for 'ontology' over 'epistemology', whatever that move might entail (see also Pedersen 2012a). Rather, the ontological turn is an attempt to reconceptualize the distinction between the two, such that questions deemed 'epistemological', which have to do with the nature of anthropological knowledge, are addressed with reference to irreducibly ontological considerations, which, it is our contention, arise from all anthropological inquiry (see also Box 4.1).

This, then, is the analogy between anthropology and animism: Insofar as perspectivism is an Amerindian model of seeing, its basic concern can be said to be epistemological. As it turns out, however, in this instance what counts as 'seeing' is also an irreducibly ontological operation. In Amerindian cosmologies, as we have discussed, different species do not see the same thing (or 'world') in different ways, but rather 'see in the *same* way ... *different* things' (1998: 478). Essentially, then, Viveiros de Castro's turn to ontology consists in experimentally and reflexively investigating what would happen to anthropology itself if it conceived of its trademark concern with difference in a similar way. What would a multinaturalist anthropology be like?

Box 4.1: Ontology in the mirror: Viveiros de Castro and Wagner

Viveiros de Castro's manner of connecting the perspectivism of the Amerindians and the ontological turn of the anthropologists strictly mirrors Wagner's connection between Daribi cultures of invention and his own anthropological invention of culture, which we outlined in Chapter 2. Like Wagner, Viveiros de Castro uses ethnographic materials to recast the anthropological distinction between nature and culture. However, while Wagner's argument reworks this distinction from the culture end, Viveiros de Castro's starts from the end of nature. If for Wagner what stands most obstructively in the way of making sense of Daribi manners of invention is the idea that culture must consist in a set of conventions, for Viveiros de Castro the prime sticking point for conceptualizing Amerindian animism is the idea that nature must be uniform and that differences in what different species 'see' can only be a matter of the varying perspectives they take upon it. Accordingly, while Wagner's reflexive reconceptualization produces a novel way of thinking about culture, namely as a process of invention, Viveiros de Castro's argument issues in a radical reconceptualization of nature, namely as a field of multiplicity: a 'multinature'.

But now consider the twist. Wagner's concept of invention does have ontological consequences, since, as we showed, it effectively turns epistemic representations into ontological operations of conceptual transformation. But as we also saw, Wagner himself hardly uses the language of ontology to explain this – we had to draw these metaphysical consequences out of his argument ourselves. You don't need to think about ontology in order to articulate the idea of culture as invention (though, importantly, if you want to you can). By contrast, if your concern is with recasting the idea of nature, you very much need to talk about ontology. 'Nature', after all, is understood conventionally as the domain of what 'is' – what exists – and contrasted with culture as the realm of what people think or imagine exists. So any attempt to reconceptualize the notion of nature is *ipso facto* an intervention in questions of ontology. In contrast to Wagner's theorization of culture as invention, in which questions of ontology remain submerged, Viveiros de Castro's conception of multinaturalism *cannot but* put the notion of ontology at the centre of anthropological debate.

Posing the question in this way allows us to address head on one of the prime misconceptions that Viveiros de Castro's work, and indeed the ontological turn as a whole, has attracted in the critical literature, namely the tendency, in one way or other, to imagine the key idea of ontological difference as a new version of the familiar anthropological notion of cultural difference – 'ontology', as the motion for a much-cited debate on the topic had it, as 'just another word for culture' (Venkatesan et al. 2010). With reference to the analogy with Amerindian perspectivism, it is not difficult to see how such a misconception might arise. If the multinaturalism of the Amerindians consists in different species seeing in the same way different things, then in its anthropological version multinaturalism must surely (so the misconception goes) consist in different peoples (societies, groups, or whatever the relevant unit may be in each case) doing something analogous. Thus Viveiros de Castro's oft used phrase, 'multiple ontologies' (e.g. 2012), would be read as a replacement for more familiar anthropological devices for designating peoples' manners of 'seeing the world', such as culture (or collective representations, discourse, habitus, social constructions, worldview or whatever). We would arrive, then, at a particularly crass version of the theoretical approach that we in the first chapter of this book referred to as 'deep ontologies' – a bizarrely shallow one at that, not to mention politically problematic.

Given how pervasive this misconstrual has been in the recent literature (Harris & Robb 2012; Laidlaw 2012; Pina-Cabral 2014; Bessire & Bond 2014; Vigh & Sausdal 2014), one may fairly wonder whether Viveiros de Castro and his followers may bear some of the responsibility for it. Certainly, loose talk of 'multiple ontologies' or 'indigenous ontologies', often territorialized as 'Amerindian', 'Melanesian', 'Mongolian', 'Western' and so on, and sometimes even tagged with the preposterously reifying image of 'many worlds' (e.g. Henare et al. 2007), has been no help. Still, we would note that, understandable as they are, such geo-cultural territorializations of 'ontologies' (and note the tell-tale noun-form) end up rendering Viveiros de Castro's anthropological transposition of his argument on perspectivism altogether inconsistent. As we saw already in Chapter 1 in relation to the 'deep ontologies' approach adopted by Michael Scott, Philippe Descola and others, the idea that ontological differences can be projected onto the world, even as they multiply it by the units of analysis anthropologists habitually think of as cultures, societies and so on, ratifies the basic premise that perspectivism concertedly denies: namely, that perspectives differ from each other with respect to a world (now a 'multiple' one, to be sure), located out there, somewhere beyond them. If one leaves this basic idea intact, multinaturalism turns into a sort of conceptual car-crash, with each perspective (here still conceived as a new placeholder for something equivalent to what we've always called 'culture') projecting a 'world' of its own. In this way, the kinds of differences anthropologists habitually imagine as cultures are hardened and pluralized into a sort of perverse shackle of isolated alterity, or even an 'ontological apartheid' (Laidlaw 2012), and subjected to the usual charges of essentialism, exoticism, reification, domination and all the other insults anthropologists consign themselves to trading for as long as they persist in thinking of their task as that of 'representing' (taking a perspective on) the 'world' (be it the human, social, cultural or indeed natural world).

Viveiros de Castro's analogy between anthropology and perspectivism works altogether differently. The multinaturalism of his anthropology resides in 'ontologizing' not the differences anthropology has posited between cultures (or other such territorialized placeholders, be they 'narratives', 'imaginaries' or what-have-you). Rather, what for him constitutes the key site of ontological difference are the divergences of perspective that are internal to the *activity of anthropology itself*, namely the perspective of the anthropologist and that of the people he studies, or as Viveiros de Castro puts it figuratively, 'the native'. Now, before rushing to conclusions about the problems conjured by this term and its questionable historical and political baggage, it is important to note that, for Viveiros de Castro, the difference between 'anthropologists' and 'natives' lies not, as he puts it, 'in the so-called nature of things' (2013: 475). On the contrary, this difference is nothing other than a feature of anthropological inquiry, understood as a 'language game' imbued with its own 'ground rules' (2013: 475):

The 'anthropologist' is a person whose discourse concerns the discourse of a 'native'. The native need not be overly savage, traditionalist nor, indeed, native to the place where the anthropologist finds him. The anthropologist, on his part, need not be excessively civilized, modernist, or even foreign to the people his discourse concerns. The discourses in question (and particularly that of the native) are not necessarily texts, but rather may include all types of meaning practice. What is essential, however, is that the discourse of the anthropologist ... establishes a certain relation with that of the native ... This relation is one of meaning or, when the anthropologist's discourse aspires to be Scientific, a relation of knowledge. (2013: 473–4)

So, as with Amerindian perspectivism, anthropological multinaturalism does not pertain to the relationships between transcendent objects of description located somewhere in the world 'out there'. It pertains to immanent relations of difference that are internal to the language of anthropological description itself, namely the differences between perspectives as they engage each other as subjects – the deictic relationships of 'I' and 'you' that hold between an anthropologist and his interlocutor(s) during fieldwork, or even between the familiar (and proverbially evil) anthropological duo, 'us' and 'them', provided the plural pronouns in question are taken to designate, not anthropologists and natives as exemplars of their respective 'cultures', but rather the contrasting but mutually constitutive positions entailed by the very activity of anthropological inquiry. Rather like as with actors and audiences in the theatre according to Peter Brook (1968), to do anthropology all you need is a set of people and someone doing a study of them.[7]

---

[7] This, if you like, is the anthropological cogito, put in the starkest possible terms: *I think, therefore you are* (*qua* subject of my inquiry, of course). Indeed, the formula could also be reversed: you are, *qua* subject of my inquiry, therefore I think. Which is to say, you, the native, force me, the inquirer, to shift my thinking in new ways,

The respective perspectives of anthropologist and native are enunciated as different, then, purely due to their constitutively divergent disposition with respect to the very activity of anthropological inquiry – nothing more, nothing less. Viveiros de Castro's ontological turn consists in saying that just this intrinsically anthropological divergence should be conceived as ontological as opposed to merely epistemological. In this way, the perspectivist slogan – 'seeing in the same way things that are actually different' – is transposed onto the structure of anthropological inquiry itself. To illustrate how this works in practice we may adopt Malinowski's language of anthropological and native 'points of view' (Malinowski 1961; cf. Holbraad 2013a; Viveiros de Castro 2014: 77–8) and use it to run the idea of perspectivism through the example of Mauss's account of Maori gifts, with which we opened the Introduction of this book.

The standard way to think of Mauss's famous Maori example would be to say that the native 'sees' a spirit in what the anthropologist 'sees' merely as an object. One thing, two viewpoints on it – one nature, two cultures. By contrast, on Viveiros de Castro's perspectivist account, the divergence between the two perspectives consists, not in two ways of 'seeing' the gift, but rather in two different ways in conceiving what the gift *is*. And note that the analogy with Amerindian perspectivism is strict. Anthropologist and native 'see in the same way' inasmuch as they both 'see' gifts. After all, the anthropologist starts off knowing what a gift is (indeed, so does the imaginary student of our opening illustration in the Introduction) and, as per the anthropological description, so do 'the Maori'. Their respective perspectives, one might say, project 'gifts' in the same sense as the respective perspectives of humans and jaguars in Amazonia project 'beer'. But then, just as with the latter, the problem

as per the ontological turn's conception of anthropological inquiry as an experimental procedure of reflexive reconceptualization in the face of ethnographic contingency.

is that what a gift is in either case 'is actually different', as per the perspectivist formula. A 'mere' object (for the anthropologist) and an object imbued with a spirit (for the native) are *two different things*. So, the divergence between the perspectives upon which the very notion of anthropological inquiry is premised is irreducibly ontological (as opposed to just epistemological, although it is of course that too).

Since this is really the heart of Viveiros de Castro's argument – indeed, it is the heart of the ontological turn – let us dwell for just a little longer on the standard misconception that all this could be understood as an attempt to reify as ontological the differences between cultures. For the misunderstanding is itself instructive. Its chief mistake is that it swallows up Viveiros de Castro's reflexive concern with the relationship between 'anthropologist' and 'native', as the constitutive perspectives of anthropological inquiry, by glossing it as an exemplar of the ontological differences between the 'wider' cultures to which these perspectives supposedly must belong. Thus the set-up of anthropological inquiry becomes the dependent variable, if you like, of an encompassing model of cultural divergences that, for its part, takes the role of independent variable.[8] By contrast, Viveiros de Castro's idea is that, within the economy of anthropological inquiry, the independent variable must be the structure inherent to that very activity, namely the mutually defining perspectives of the anthropologist and the native, as per the 'ground rules' laid out previously. But then the notion of 'cultural divergence', along with the whole conceptual infrastructure that constitutes the regime of multiculturalism, gets relegated to the position of dependent variable. It becomes but one of the forms that can be projected within the economy of anthropological analysis, be

---

[8] The effect of this, as we have seen, is to break up the immanent relationship between the anthropologist's and the native's perspective as defining tropes of the anthropological language game, and render it as a divergence between mutually transcendent cultures, and between flesh-and-blood 'anthropologists' and 'natives', assumed to exist 'out there' in the world.

that from the perspective of the anthropologist (the standard image against which multinaturalism is pitted would be the example) or from that of the native (e.g. treating 'the Moderns' as an ethnographic object, as Latour does).

Crucially, this conceptual regime must vary as it enters into anthropological analysis. What happens to the standard anthropological regime of multiculturalism when it tries to include in its scope – when it tries to 'see' – a description of a native regime that cuts against it? Answer # 1, precipitated by the regime of invention of the Daribi: obviation (see Chapter 2). Answer # 2, precipitated by the regimes of anthropological comparison when they are compared to themselves as in Strathern's *Partial Connections*: self-scaling relations (see Chapter 3). Answer # 3, precipitated by the regime of Amerindian animism: multinaturalism. And so on. This, in other words, is how we get to the ontological turn: by performing a Wagner-style figure-ground reversal between anthropology and the world, in the following sense. Anthropologists have assumed that they must take for granted what the world is (e.g. it is a place made up of one nature and many cultures) in order then to find ways to fit into it the activity of anthropology (e.g. should it take nature as cultural or culture as natural?). Viveiros de Castro, and the ontological turn as whole, invites us *to take for granted the structure of the activity of anthropology, so as to be able to ask what the world might be.* And the latter question resolves itself in a series of further questions regarding how to conceptualize the various, and by this account varying, constituents of the world: not only natures, cultures, persons, things, animals, spirits, kinship and gifts, but also money, migration, the state, security regimes, infrastructure, Big Data and the host of other such anthropological topics that, as we pointed out in the Introduction, are dealt with in the contemporary literature adopting the ontological turn as their approach.

Box 4.2: Viveiros de Castro, Deleuze and anthropology

Viveiros de Castro's tendency to couch his decolonizing arguments in the conceptual language of Deleuze may seem as something of a performative contradiction. Is it consistent to claim that native thought can 'put a bomb' (Latour 2009) under the global intellectual dominance of occidental philosophy and, in the same breath, proclaim the Deleuzian credentials of that move, couching one's very account of Amazonian social life in the language of rhizomes, virtualities, moleculars, multiplicities and so on? Taking also into account his unabashed love of Lévi-Strauss, one may wonder where exactly the geoconceptual, if not geophilosophical (Deleuze & Guattari 1994), centre of Viveiros de Castro's anthropology is to be found – is it the lowlands of Amazonia, or is it Paris (Turner 2009; Vigh & Sausdal 2014)? At any rate, doesn't his sheer, 1968-fuelled enthusiasm for Parisian radicalism tend to detract from the clarity of his otherwise relentlessly anthropological message (cf. Laidlaw 2012)? At times Viveiros de Castro even comes close to sounding like an exponent of what in Chapter 1 we called the 'alternative ontology' tendency in anthropology, finding Deleuzian truth in indigenous universes.

It is easy to see why one might be left with this impression. Indeed, a similar observation might be made with respect to Marilyn Strathern and Roy Wagner's work on Melanesia. At any rate, as several commentators have pointed out (e.g. Bennett & Frow 2008: 38; Jensen & Rödje 2009; Dulley 2015; see also Morris 2007), so-called New Melanesian Ethnography displays striking resemblances to core theoretical ideas associated with poststructuralist philosophers, including Derrida as well as Deleuze. Still, with regard to his own work, Viveiros de Castro has sought to make clear that his appeal to Deleuze is not as a philosophical patron for multinaturalist anthropology, much less for Amerindian perspectivism (2014: 91–106 and *passim*). For one thing, to claim so would be to court tautology, since a great deal of the conceptual production of Deleuze (and Guattari) in particular can be traced back to Amazonia, Melanesia and other indigenous sources, via the influence exerted on Deleuze not only by Lévi-Strauss himself, but also Gregory Bateson, Marcel Griaule and other anthropologists (Jensen & Rödje 2009; Viveiros de Castro 2009). Furthermore, in Viveiros de Castro's own work, the flow of meaning can be imagined both ways, since one could just as well say that in it Deleuze is read 'through' Amerindian conceptions, even as Amazonia is read through Deleuzian ones. At root, for Viveiros de Castro, the immanent connection between the two conceptual regimes comes down to a *political alliance*. In Deleuze, for whom the philosophical love of the concept acts to keep thought 'radically, anarchistically plural' (Skafish 2014: 18), Viveiros de Castro is looking not for a philosophical guarantor, but rather for a comrade in arms.

This is what Viveiros de Castro refers to as 'an anthropological concept of the concept' (e.g. 2013: 483–8; 2014: 80 and *passim*). For him, this is important partly as a way of connecting anthropology with Deleuze's philosophy, which Deleuze famously saw as an activity devoted to the creation of concepts (Deleuze & Guattari 1994; see also Box 4.2). And indeed, such an alliance captures nicely the intellectual excitement that this release of conceptual energy represents for anthropology (see also Jensen & Rödje 2009). Pursuing the analogy, he invites us to imagine an anthropology that could take 'indigenous conceptions as being of a kind with the *cogito* or the monad' (Viveiros de Castro 2013: 487):

We could thus say ... that the Melanesian concept of the 'dividual' person (Strathern 1988) is as imaginative as Locke's possessive individualism; that understanding the 'Amerindian philosophy of chieftainship' (Clastres 1974) is as important as commenting on Hegel's doctrine of the State; that Māori cosmology is equivalent to the Eleatic paradoxes and Kantian antinomies (Schrempp 1992); that Amazonian perspectivism presents a philosophical challenge of the same order as Leibniz's system ... And when it comes to what matters most in any given philosophical elaboration, namely its capacity to create new concepts, then without any desire to take the place of philosophy, anthropology can be recognized as a formidable philosophical *instrument* in its own right, capable of broadening a little the otherwise rather ethnocentric horizons of our philosophy. (Viveiros de Castro 2013: 487–8; references and emphasis in the original)

Thus the paradigm shift (for that is what it is, in the strictest sense) of moving away from the standard anthropological obsession with measuring up to the epistemic standards of science (e.g. as 'social scientists' concerned with 'representations') and towards a traditionally philosophical 'love of the concept' is not *just* an attempt to release the discipline's potential for conceptual creativity. It is also, and by the same token, an attempt to get anthropology out of its central methodological-cum-existential concern, namely the perennial question of how to describe, interpret, explain, analyse, speak to, from or for the people we study without smothering them with our own ways of thinking and acting – this being the anxiety underlying the worries with ethnocentrism, exoticism, essentialism and all the other dirty words of the discipline, with which we framed the overall argument of this book in the Introduction. For Viveiros de Castro the significance of this problem is above all political. At stake is really how seriously we, as anthropologists, are prepared to take the people whose lives we seek to elucidate. To see this, we now turn to the key methodological consequences of multinaturalist anthropology, with reference to what Viveiros de Castro's calls the 'method of controlled equivocation' (2004) – his way of taking people seriously.

Taking People Seriously

'Anthropology is that Western intellectual endeavour dedicated to taking seriously what Western intellectuals cannot ... take seriously.... Doing so demands as much sense of humour as its converse, namely not to take seriously what we "simply" cannot not take seriously' (Viveiros de Castro 2011a: 133). So goes one of Viveiros de Castro's oft-repeated calls for anthropologists to take seriously the people they study. But how does he go about this, and how is his way of doing so different from other anthropologists'? After all, one would have to look hard to find an anthropologists who would *not* see it as their mission to take their interlocutors seriously – an aspiration for the discipline that is as old as Malinowski. The 'method of controlled equivocation' is Viveiros de Castro's answer to these questions.

'Equivocation', Viveiros de Castro suggests, is what happens when you try to translate across 'native' and 'anthropological' conceptual regimes. The word's connotations are hardly accidental: such translations are certainly equivocal and hesitant (as we saw in relation to Strathern), turning as they do on the inherent multiplicity of meaning (this being the technical philosophical sense of 'equivocity'). Add to this the colloquial Romance-language sense of equivocation as 'error' and you get the gist of Viveiros de Castro's conception. If the difference between the conceptual regime of the 'anthropologist' and that of the 'native' (*again and always: understood in terms of their internal relation in the economy of anthropological inquiry*) is ontological, translations from one to the other needs must take the form of errors of a particular sort, namely, *misunderstandings*. The jaguar and the hunter both think they see manioc beer, but actually, per the perspective of either, they are looking at different things entirely. The 'native' and the 'anthropologist' both talk about gifts, but again, actually, one is talking about something (somehow) containing a spirit while the other is talking about a mere object. So if the economy of anthropological inquiry is constituted by ontological divergences, it is also an economy of misunderstandings. Within the language game that is at issue here, then, 'the anthropologist' and 'the native' do not so much disagree with each other (Gifts have spirits! No, they don't...) as constitutively *talk past each other*.

So how might translation be possible? What, if you like, is the anthropological equivalent of shamanic travel? Of course, we are too far into this book not to have a fairly clear notion of the answer. Wagner's arguments for invention, his method of obviational analysis, Strathern's relations and the 'least obvious' postplural and trans-temporal comparisons she uses them to make, as well as Viveiros de Castro's own scheme for perspectivism and multinaturalism – all these are attempts to enunciate, and in so doing also exemplify, manners in which ontological divergences identified in ethnographic inquiry can act to propel rather than inhibit anthropological analysis. Indeed, having examined each of these arguments in detail, we are now in the position to note a basic feature they all have in common. Namely, the idea of *self-transformation* – that is, the very idea Viveiros de Castro finds so amply in Lévi-Strauss and puts at the heart of his own model of perspectivism. Invention, obviation, relation, postplural abstraction, perspectivism: all are conceptualizations arrived at by transforming a set of anthropological assumptions so as to bring them into line with a body of ethnographic materials that initially contradicts them. This takes place in an inherently heuristic and thoroughly reflexive, almost self-abnegating manner: the notion of culture as a set of conventions is meticulously budged from its axiomatic position in order to give way to the semiotic of invention; the idea that persons are distinct both from each other and from the things they exchange is systematically eroded in order to produce the conceptual infrastructure of Melanesian sociality; the ontological coordinates of one-nature-many-cultures are carefully shifted, ninety degrees, to produce the conceptual regime of multinaturalism.

In Viveiros de Castro's terminology, each of these processes of transformation constitutes an equivocation in the fullest sense. They are all 'motivated' (*sensu* Wagner) by an initial 'error' – the conceptual mismatch the anthropologist diagnoses as a misunderstanding. They involve equivocity in the philosophical sense since they turn on the multiplicity of meaning – concepts are budged, eroded, shifted (as per Strathern's scalings also).

And this process is inherently precarious and piecemeal – virtuously and optimally hesitant for Strathern – since it strikes out into the unknown, rendering itself in and as its own variation. So if these are translations, Viveiros de Castro explains, they are so in a sense that is directly opposite to what we might expect from within the representational economy of multiculturalism. Articulating the notion of translation appropriate to the perspectivist regime of Amerindian animism, he writes:

[T]he aim of perspectivist translation ... is not that of finding a 'synonym' (a co-referential representation) in our [anthropological] conceptual language for the representations that [natives] use to speak about one and the same thing. Rather the aim is to avoid losing sight of the difference concealed within equivocal 'homonyms' between our language and that of [the people we study], since we and they are never talking about the same things. (2004: 7)

And then the idea is transposed to the activity of anthropology as follows:

To translate is to emphasize or potentialize the equivocation, that is, to open and widen the space imagined not to exist between the conceptual languages in contact, a space that the equivocation [here *qua* error] precisely concealed. The equivocation is not that which impedes the relation, but that which founds it and impels it; a difference in perspective. To translate is to presume that an equivocation always exists; it is to communicate with differences, instead of silencing the Other by presuming a univocality – the essential similarity – between what the Other and We are saying. (ibid.: 10)

How, then, is equivocation to be *controlled*? Of course, as per Strathern's aforementioned observation that 'what is true of what is observed is true also of the manner of observation' (2014a: 7), it is clear that the various arguments about invention, obviation, relation, comparison and so forth that we have already explored are all manners not only of *doing* controlled equivocation, but also of explaining how it works. (In that sense our own attempt to model them as such by way of this book's 'exposition' is also one big exercise-cum-demonstration of controlled equivocation.) What Viveiros de Castro adds to this is a conceptual vocabulary that makes the ontological stakes of the exercise entirely explicit, by appeal to the philosophical – indeed Scholastic – distinction between extensional and intensional understandings of meaning. The argument is worth rehearsing, since it provides perhaps the most incisive way of expressing the essential 'move' of the ontological turn – its bottom line, so to speak.

The extension of an expression is its reference. For example, if I ask you what a jaguar is and you point one out to me by showing me, say, a photo ('there's one!'), you are giving me the meaning of 'jaguar' in terms of its extension. The intension of an expression, by contrast, comprises the criteria (i.e. the sufficient and/or necessary conditions) for determining its extension (Chalmers 2002; cf. Putnam 1975). So if instead of showing me a picture you tell me, as in a dictionary, that a jaguar is a large feline predator that lives in the Americas, you will be giving me the intension of the term. So, ringing the bells of our discussion so far, we may say that extensions pertain to the transcendent relationships between representations and the things they represent, while intensions pertain to the immanent relations between concepts that define each other's meanings. Or, in Wagner's terms, semiotics of convention on the one hand and semiotics of invention on the other. Or even: epistemological purchase versus ontological definition.

Viveiros de Castro's method of controlled equivocation, then, posits translations between native and anthropological conceptualizations as a matter of modulating intensions rather than fixing extensions. If a native person assumes (or even says, as the Ranapiri did to Elsdon Best in the famous passage of Mauss's analysis) that what both he and the anthropologist see as a gift is something that contains a spirit, then the anthropologist must ask himself reflexively: what is it about the way I define gifts that makes this native assumption appear incongruous? How do I need to change my definition in order to remove this intensional incongruity?

For example, if 'the native's' contention that the gift contains a spirit clashes with my own assumption (*qua* analyst) that it is an inanimate object, then it is my job to rethink first of all the operative distinction between spirits and objects, which seems to underwrite my conception of what a gift is. How might I modulate this distinction, altering the relational constitution of the meaning of the concepts involved (e.g. thing/spirit, object/subject, donor/recipient, alienability/inalienability)? Moreover, where does this leave the intensions of a series of corollary assumptions that will inevitably become relevant as my effort to conceptualize gifts is brought to bear on the full complexity of the ethnography of Maori gift-exchange? For example, what might 'property', 'labour', 'fame', 'honour', 'profit' or 'self-interest' mean in the context of these transactions, and are they best thought of as 'transactions' at all? Further still, what notion of 'action' is appropriate here? What counts as an 'intention', or a 'cause', or an 'effect', and how then, if at all, might an action be distinguished from an 'event'?[9]

Each of these questions helps to guide – or 'control' – the intensional modulation of anthropological conceptions in the light of the ethnographic materials to which they are exposed, transforming them piecemeal into concepts that may accord with the conceptions that these materials express. Anthropological analysis, on this account, emerges as an ever-precarious, complex and inherently experimental exercise of mutual calibration between ethnographic materials and their anthropological conceptualization – an experiment that is not just likely to 'fail', but is in fact *supposed* to do so (as also discussed in the Introduction). Indeed, lest the emphasis on conceptualization give the impression that Viveiros de Castro's method of controlled equivocation (not to mention the ontological turn more generally) consigns anthropology to some neverland of philosophical reflection and self-indulgent intellectual play, conducted at the expense of the nitty-gritty realities of which

---

[9] Readers may recognize these examples from, not just Strathern's work, in mutual dialogue with that of Nancy Munn (1986) and Roy Wagner, but also Annelise Riles (2001), James Leach (2003), Adam Reed (2004), Tony Crook (2007), Melissa Demian (2007), Andrew Moutu (2013), and Alice Street (2014) among many other of her ex-students working in different parts of Oceania.

ethnographic fieldwork is made (e.g. Bessire & Bond 2014), it is important to emphasize here that the ultimate 'control' over the process of anthropological conceptualization that lies at the heart of Viveiros de Castro's method is exerted precisely by the ethnographic descriptions, in all their depth and complexity. Our use of the Maussian 'take home message' about gifts containing spirits as an example is just that: an illustration pared down to a bare moment of a-ha! insight that may be familiar to readers from their first-year courses as students. We know *of course* that ethnographies of Maori and other comparable regimes of exchange are infinitely more complex and therefore also less blatantly 'exotic' than that, and that these giftlike moments of insight are distilled out of this complexity. It is to this depth and complexity that anthropological conceptualizations are answerable: our equivocal conceptions stand or fall by the degree of light they shed upon them. If we may offer an argument from authority here, Strathern's *The Gender of Gift* (1988) can serve as an example of the level of ethnographic breadth, depth and sophistication to which the controlled equivocations of the ontological turn can – must – aspire.

As long, that is, as ethnographic accounts such as Strathern's are not taken as attempts to represent a socio-cultural reality 'out there', that stands as the transcendent benchmark against which the empirical purchase of such concepts as the 'dividual' could be measured. That, as Viveiros de Castro puts it (2013: 477), is the 'other game' – the one in which anthropological translation is imagined as a matter of getting right the referential extension of the terms in which it is conducted. Indeed, with great polemical force, he draws out the consequences of the intensional modulation on which the method of controlled equivocation is based by contrasting it with just that other game. Since this is the contrast on which the epistemological-cum-political question of 'taking seriously' native conceptions hangs, we may pursue it in some detail.

Extensional translation, Viveiros de Castro argues, is the most familiar anthropological strategy. You take your ethnographic material and describe it using the closest synonyms you can find for it in your own conceptual vocabulary. To stay with the example, when the Maori exchange objects – giving one ceremonially, expecting one in return, – you may say they are exchanging gifts. 'Gift', let us assume, is the term whose extension comes the closest to coinciding with the exchanges you wish descriptively to identify. But of course you soon find out that there are limits to the synonymy, and this is when things get interesting. For while 'gift', as you understand it, refers to certain objects of exchange, the Maori (according to Mauss) also hold that they contain spirits. So, given that 'objects' as we ordinarily understand them are the kinds of things that do *not* contain spirits, the Maori seem here to be holding a pretty odd view on gifts... But note that this oddness corresponds exactly to the degree in which it *exceeds* the extensional limits of the notion of 'gift' the anthropologist takes as his point of departure. And this is where anthropological analysis of the familiar, 'cross-cultural translation' type kicks in: How might we 'explain' the odd Maori idea that gifts contain spirits? Does it help with the smooth functioning of society? Is it a local expression of some universal economic process? Or perhaps of universal cognitive processes? Alternatively, is it a metaphor of some sort, symbolic of something or other? At any rate, might we not try to make sense of it in relation to other local ideas and practices, interpreting it in its cultural context?

This, of course, is the stuff of heated and in many ways defining theoretical debates within anthropology, and generation after generation of students have been taught the history of the discipline as a series of choices between the different options these debates mark out. However, for Viveiros de Castro, what all these otherwise contrasting theoretical positions have in common is the basic move of projecting onto the native a sense of mystification, not to say 'error', that should properly belong *to the anthropologist* (and here, of course, we are talking of the native not as a deictic position within the economy of anthropological inquiry, but rather as an objectified flesh-and-blood person out there, whom the anthropologist – also now a flesh-and-blood subject – studies empirically). By deciding a priori that we know *what* the native is talking about (e.g. 'the Ranapiri is talking about gifts') the divergence that we detect a posteriori in what he says must take the form of a *deviation* from what we take to be obvious (e.g. that gifts are just objects exchanged between people).

Of course, what one should do with this deviation is the prime question for anthropological theory, thus construed. You can choose to affirm it as an error or, as it was tellingly debated for decades, as a form of 'apparently irrational belief' (e.g. Sperber 1985), and then set yourself on explaining why people might nevertheless be wont (so often!) to commit it. Or you may choose to be more liberal and relativist, conceding that there's no way of knowing who's right and who's wrong in these things, since the views of the anthropologist and his or her interlocutors are both socially constructed, and so on. Still, none of this takes away from the fact that, once the distance between your position and that of your ethnographic interlocutors is con-ceived as a matter of extensional divergence, the anthropological challenge is *a fortiori* set up as a matter of accounting for (explaining, interpreting, contextualizing) the divergence of what your interlocutors (supposedly) 'believe' from what you take to be the case. Indeed, the tell-tale sign of the epistemological asymmetry of this way of setting up the problem is in the word anthropologists so naturally use to gloss the views of the people they study, namely 'beliefs' (Viveiros de Castro 2013: 488–96; Latour 1993, 2010). Belief, after all, is what we call a view to which we are reluctant to commit – a view we can't quite bring ourselves to take entirely seriously.

This is what Viveiros de Castro has in mind, then, when he speaks of controlled equivocation as a manner of 'taking people seriously'. Indeed, as we shall discuss presently, for him the bottom line in choosing between intensional and extensional renditions of anthro-pology – 'there are two incompatible ways of conceiving anthropol-ogy, and one has to choose between them', he writes[10] (2013: 477) – is

[10] In the course of developing a conceptualization of truth appropriate to the con-ceptual shifts of the ontological turn, Holbraad has taken issue with Viveiros de

ethical and, ultimately, political: not to take others seriously is to silence them and in that way to dominate them. But here we should emphasize that such an explicitly ethical and political argument is underwritten also by a more implicit methodological one – a real clincher, we would suggest, that tends to be overlooked in Viveiros de Castro's polemics. And this is that anthropologists' focus on the so-called 'beliefs' of the people they study only serves to gloss over a basic *failure of description on their own part*. Deciding by fiat that when Maori speak about things they exchange, they speak about what you have in mind when you think of 'gifts', only, on top of that, they 'believe' that these gifts contain spirits (and what a cumbersome manner of 'saving' your concept of the gift from the natives this is!), is not just a manner of projecting a form of error onto the people you are meant to be trying to understand. It is above all a failure to describe what these people are saying, doing, or assuming, as the case may be. In anthropology we are so accustomed to this kind of belief-talk that we seem to have learnt to overlook the ambiguities, confusions, incongruities, contradictions, and sometimes downright incoherence that it so characteristically involves. Things that have spirits, pasts that are present, societies that are bigger than the sum of their parts, and so on, in near-Dadaist ethnographic wonder (cf. Scott 2013a; 2014). Evocative as they may sound, however, the one thing that

Castro's tendency to pit representation (or extension) and equivocation/concep-tualization (or intension) against each other in this either/or way (2012: 49–53, 255–8; 2013). The litmus test for the cogency of anthropological conceptualiza-tion, Holbraad suggests, is representational. If one modulates one's concepts in order to remove one's initial inability to describe one's ethnographic materials, then one can only know that one's modulations are cogent when one reaches a point at which one *can* describe the material in question. Description, here, is an act of representation in the ordinary sense, and is an indispensable ingredient or phase in the process of conceptualization, marking its beginning and its end. So rather than favouring conceptualization at the expense of representation alto-gether, Holbraad suggests, the ontological turn performs a figure/ground reversal between them. Such a reversal is strictly analogous to the one Wagner's semiotic of invention performs with inventions and conventions.

such quintessentially anthropological statements have in common is a failure to express a clear meaning: they are all, strictly speaking, contradictions in terms. If we know what a thing is at all, we know that it has no spirit. The whole point about the past is that it is *not* present any more – it's gone! And the whole *just is*, by its very definition, the sum of its parts. To say that people 'believe' these things is strictly equivalent to saying they 'believe' that 2 + 2 = 3 – a statement lacking a conceptual correlate.

When it comes to taking people seriously, then, the problem is not with people's beliefs, but with our own inability as anthropologists sensibly to convey what these people say and do. As we have seen, doing so may involve *modulating* the conceptual repertoire on which the anthropologist relies, so as to be able to *arrive* at sensible descriptions of the ethnographic objects in question – that being the purpose of the ontological turn.

Conceptual Self-determination

Having sought to clarify the principal tenets of Viveiros de Castro's anthropological project, we are now in a position to understand the full significance of his somewhat pompous-sounding call for the 'conceptual self-determination of the world's peoples' (2003), with which we started this exposition. There is, of course, a strong dose of irony in the statement. The call to 'self-determination', not least in the name of 'peoples', has more than a tinge of Enlightenment high spirits about it, and these are precisely the kinds of traditions of political thinking under which a multinaturalist anthropology puts a bomb, in Latour's memorable expression. Still, the irony is the point (Skafish 2014: 11). The difference it makes to call for the *conceptual* self-determination of the peoples is that each of the notions that this reflexive sentence enunciates is put up for wholesale redefinition. What 'people' might be, in what their 'self-determination' might consist, and what, indeed, a 'concept' might become, are precisely the quintessentially ontological questions the sentence poses. By way of conclusion, then, we would like to dwell a little longer at the threshold – indeed the 'limit' – of these questions in order to clarify the political stakes of Viveiros de Castro's anthropological project.

In particular, we would like to address a sense of irritation, often expressed by critics (Ramos 2012; Bessire & Bond 2014; Salmond 2014; Graeber 2015), with Viveiros de Castro's apparently self-aggrandizing political pretensions. For it may be fair to wonder what kind of political emancipation is on offer here. How exactly do 'the peoples of the world' benefit from Viveiros de Castro's supremely smart (granted) forms of conceptual experimentation and theoretical creativity, which seem to be so rarefied that even many of his anthropological colleagues persistently misunderstand them? What help, in the end, is Viveiros de Castro's perspectivism to the 'Indians'[11] he so likes to talk about? In a recent manifesto-like pronouncement on the politics of the ontological turn, Viveiros de Castro sought to take the sting out of this line of attack, by returning to basic political premises:

The first (unproductive) misunderstanding that should be dispelled [in relation to the claim about conceptual self-determination of the peoples] is the idea that this is equivalent to fighting for indigenous peoples' rights in the face of the world powers. One does not need much anthropology to join the struggle against the political domination and economic exploitation of indigenous peoples across the world. It should be enough to be a tolerably

---

[11] A charge that often rides on the idea that Viveiros de Castro's project of conceptual decolonization is not sufficiently 'political' is that it exoticizes its subject matter, thus marking anthropology's latest 'return of the primitive' (Bessire & Bond 2014: 442). Eduardo Kohn's commentary on this line of critique puts the rationale of Viveiros de Castro's own position well: 'Anthropology surely has a nostalgic relation to the kinds of alterity that certain historical forces (which have also played a role in creating our field) have destroyed. To recognize this is one thing. It is quite another to say that for this reason there is no longer any conceptual space "alter" to the logic of this kind of domination. For this would be the final act of colonization, one that would subject the possibility of something else, located in other lived worlds, human and otherwise, to a far more permanent death' (2015: 320).

informed and reasonably decent person. Conversely, no amount of anthropological relativism and old-hand professional scepticism can serve as an
excuse for not joining that struggle. (in Holbraad, Pedersen & Viveiros de
Castro 2014)[12]

The rejoinder makes best sense in terms of the foregoing (and to Viveiros
de Castro absolutely central) distinction between the internal economy
of anthropological inquiry, on the one hand, and the world at large as we
know and live in it, on the other. Fighting for indigenous people's rights,
identifying and struggling against injustices, holding the powerful to
account – these are activities of the world we live in. As it happens,
Viveiros de Castro himself is particularly involved in them, not least as
an activist for indigenous peoples' rights in Brazil. But while this may
make him a good person, as it were, it doesn't automatically make him
a better anthropologist. Indeed, to try to define the political stakes of
anthropological inquiry by imagining for it a role within this image of
the world – a world already conceived and fully populated with peoples,
nation-states and world institutions, unequal distributions of power,
exploitative economic arrangements, and replete with human suffering –
is to ratify the conceptual premises of just that world, trumping with it the
radical potential inherent in anthropology as a mode of inquiry, namely
that of imagining, and conceptually elaborating, ethnographically
motivated alternatives to it. The goal with which Viveiros de Castro tasks
anthropology is indeed more modest than that of *changing* that world.
Anthropology cannot, for example, roll back the forces of colonialism
and postcolonialism – surely that takes political activism (indeed
action) of an altogether different order of force and scale. What it can
do, however, is operate in that direction in its own immediate ambit,
namely the economy of anthropological inquiry itself, to promote 'the

---

[12] The paragraph cited here was part of Viveiros de Castro's contribution to this
co-authored text. We shall be drawing on the parts we wrote ourselves in the
Conclusion to this book.

permanent decolonization of thinking', as Viveiros de Castro notoriously puts it (e.g. 2003; 2009). To limit oneself to the question of how best to think is not to renege on one's political responsibilities. To the contrary, it is to *take* political responsibility for the one field of action for which anthropologists, by definition, are *actually* responsible, namely their own.

If anything, one may wonder whether Viveiros de Castro should rather be held accountable for not taking his argument far *enough*. For, even making allowances for his ironic style, the relationship between the form of his political motivations, which frame his anthropological project, and the constitutively equivocal content of that project is never quite clear. In particular, one could fairly ask why the very notion of politics, let alone of the nature of one's particular political stance (e.g. unwavering support for indigenous peoples, permanent hostility to the State), is excluded from the anthropological economy of equivocations (Candea 2011a, 2011b, 2014; Ricard 2014). If, anthropologically speaking, we can (must) desist from deciding in advance what, say, 'self-determination', 'peoples' or 'world' might be, then surely we should do the same with the very notion of 'politics', which motivates this endeavour. That would of course be also a major risk. It would be a risk of the same order, perhaps, as the one Strathern runs when she defines (anthropological) feminism, not as struggle against the exploitation of women, but as a relentlessly reflexive investment in non-completeness as a form of (feminist) anthropological action.

In the Conclusion to this book we shall take up these issues again in addressing the political implications, as well as the critical potential, of anthropology's ontological turn as a whole. Before getting to that, however, and having now completed our intellectual genealogy of the ontological turn, we devote the final two chapters to exploring some of the ways in which this manner of thinking could be critically extended in directions not necessarily precipitated, or indeed endorsed, by the three primary figures of this particular theoretical linage, Wagner, Strathern and Viveiros de Castro. Accordingly, the next chapter is devoted to a body of literature developed by a mostly younger generation of scholars, including ourselves, to explore the purchase of the ontological turn on anthropological engagements with 'things' and their material properties. In the final chapter, in relation to what we call a 'post-relational' move, we extend the question about where the ontological turn might go next by asking what happens when the experimental reflexivity of this approach is turned in on itself, to reconceptualize the very idea of the relation, which, as we have seen in such detail up to now, lies at the very heart of the ontological turn as an approach to anthropological analysis.

FIVE

∾

Things as Concepts

What purchase does the ontological turn have on the study of what anthropologists and archaeologists call 'material culture' – that is, material objects, artefacts or, simply, 'things'? Conversely, how might the study of things extend the line of thinking that the ontological turn develops, allowing its argument to encompass more than anthropologists' traditional (and defining) focus on human beings and their social and cultural comportment? Might things, like people, provide a vantage from which to transform the infrastructure of anthropological thinking? Indeed, can one take things as 'seriously' as people? What might that mean, if anything, and what would it entail?

This chapter addresses these questions with reference to a growing body of literature that is located at the interface between the ontological turn in anthropology and debates in social theory more broadly, concerning the role of material objects and other non-human entities, which have been conducted under labels such as 'posthumanism', 'thing theory', 'new materialism', Object-Oriented-Ontology or, indeed, the 'material turn' (e.g. Barad 2007; Bryant 2011; Bryant, Srnicek & Harman 2011; Bogost 2012; Morton 2013). Within anthropology itself, the book that has dealt most explicitly with these questions is the edited volume *Thinking Through Things: Theorizing Artefacts Ethnographically* (2007), in which a group of younger scholars (including ourselves) sought to transpose some of the insights of, among others, Wagner, Strathern and Viveiros de Castro onto the anthropological study of material culture. As the editors of *Thinking Through Things* made clear in their manifesto-like introduction,[1] a number of the chapters in the volume place themselves squarely on the trajectory of the ontological turn. In doing so, their attempt to 'do' the ontological turn with things takes the debate about material culture in anthropology in a decidedly 'posthumanist' direction, by seeking systematically to displace such cardinally 'humanist' or 'modern' ontological axioms as the distinction between humans and non-humans, culture and nature (indeed culture and materiality also), representations and world, the spiritual and the material and, most drastically perhaps, concepts and things.

In this chapter we develop this argument further by asking how far such a posthumanist approach might allow things to have the same kinds of reflexive effects on anthropological conceptualization as people do. In other words, rather than focusing on 'taking seriously' the people for whom things may be important, which is the promise of the ontological turn with which Viveiros de Castro and other anthropologists have mostly been concerned, the prospect here is of taking seriously something akin to 'the things themselves'. Allowing things, that is, to make a difference *as things* to the way we may think of them – to help to dictate their own terms of engagement, becoming, so to speak, their own 'thing-theorists' – by virtue of the characteristics that make them most thing-like, namely what, entirely crudely for now, we may call their 'material properties'. As we shall show through a critique of the Introduction to *Thinking Through Things*, such a move has the potential drastically to raise the status of things within the economy of anthropological analysis.

---

[1] To our knowledge, the first time the tag 'ontological turn' was used with reference to this line of anthropological thinking was in this text, where the work of Wagner, Strathern and Viveiros de Castro (along with work of Alfred Gell and to some extent also Latour and Ingold) was branded as a 'quiet revolution' within the discipline (Henare et al. 2007: 7).

On the face of it, such an argument would seem to sit uncomfort-ably with the ontological turn's constitutively critical stance to just such seemingly naive humanist ways of talking. What could such straight-laced reference to 'material properties' be, other than a hook-and-sling capitulation to the 'modern regime's' distinction between the material and the non-material? To show that our argument about things involves anything but such a capitulation, we begin by reviewing in some detail the anthropological literature on materiality, where the ontological boundaries of what might constitute a 'thing' in the first place are very much at stake.

From Humanism to Posthumanism

The development of the literature on materiality in social theory over the past few decades could be plotted as a trajectory of increasingly (self-consciously) 'radical' attempts to dislodge, or even erase, the line that divides things from people. As an illustration, consider the shift from proposing that things acquire 'biographies' and a 'social life' of their own through their complex involvement in the lives of the people who engage with them (Appadurai 1986), to saying that the very distinction between people and things (or humans and non-humans) should be eliminated from the way we think about such engagements (Latour 1993; Pinney 2005). Or consider the difference between suggesting that people and things emerge out of each other dialectically (Miller 1987; 2005) and claiming that in certain contexts they are best conceived as being analogical versions of each other (Strathern 1988; 1990).

Such differences can be conceived as two broad stages on the axis of radicalism, which, following Haraway (1991; cf. Webmoor & Witmore 2009), we may tag as 'humanist' and 'posthumanist', respectively. Humanist approaches, then, seek to raise the status of things in anthro-pological analysis – to 'emancipate' them, even (see Latour 1999; Olsen 2003; Fowles 2008; 2010) – *in terms* of the ontological division between humans and things, which they otherwise leave intact. Posthumanist approaches, by contrast, do so by *going beyond* this distinction, refiguring it or even cancelling it in different ways. So, the contrast between humanism and posthumanism corresponds to two different ways of 'taking things seriously'. Humanist approaches do so by showing how closely things are intertwined with the lives of humans, allowing some of the light of what it is to be human to shine on them too. Posthumanist approaches seek, rather, to enhance the status of things, not by associating them with humans, but in their own right, showing how, suitably reconceived, things can radiate light of their own. Yet, as we shall see, there are some important question marks about the way posthumanist approaches have gone about showing this. So let us begin by exploring the distinction between humanism and posthumanism with reference to parts of the literature that have been most influential in the debates on materiality in anthropology in particular.

Daniel Miller's introduction to his edited volume *Materiality* (2005) presents a consummate and highly influential example of the humanist approach, and allows us to illustrate how things can be taken seriously 'by association', in this case by establishing their irreducible, constitutive role in the lives of humans. To be sure, with a clear nod towards posthumanist ontological revisionism, Miller does underline the importance to anthropological discussions of materiality of 'philosophical resolution[s] to the problematic dualism between people and things' (2005: 41), and includes as an example his own preference for theorizing the relationship between people and things in terms of a Hegelian dialectical process of objectification (1987; 2005: 9). Nevertheless, for Miller the job of anthropologists cannot be simply (or complexly) to reinvent such philosophical wheels, not least because the people they study ethnographically so often have a much more common-sense understanding of things, including all sorts of ways of distinguishing them from people, or for that matter from spirits (to recall our earlier references to Maussian conceptualization of gifts). Ultimately, Miller maintains, the role of an anthropology committed to reflecting ethnographically on the world in which we live, and to theorizing what it is to be human, must be to recognize and 'respect' (2005: 38) material objects and the implicit as well as explicit ways in which they give form to people's lives. Anthropology's aim, through strategic combinations of dualism-busting philosophical models and ethnographic sensitivity and empathy, must be to show the myriad ways in which, as Miller puts it memorably, 'the things people make, make people' (ibid.).

It is perhaps not entirely clear how Miller squares the circle of the contrasting demands of a philosophical impulse to overcome dualism and an anthropological one to dwell on the myriad forms in which it may play itself out ethnographically (though see Miller 2005: 43–6). What is abundantly clear, however, is that his own heart lies with the messiness of the ethnography, and what he calls the 'vulgar' study of 'the way the specific character of people emerges from their interaction with the material world through practice' (2007: 26). If he is interested in emancipating the thing from the 'tyranny of the subject', as he puts it (2005: 45), then that is because doing so leads to a more profound understanding of what it is to be human. Miller's programme for material culture studies, which he conceives as an eclectically interdisciplinary field of its own, seeks to displace an anthropology obsessed with the imperium of the social, but only to replace it with a better anthropology humble enough to recognize the ways in which things also so pervasively contribute to our humanity. This is exactly the kind of stance we have in mind when talking of humanism and its elevation of the status of things by association to humans.

Alfred Gell's equally influential argument about the 'agency' of artefacts, presented in his modern anthropological classic *Art and Agency* (1998), provides another example of the humanist approach, though in a different and less direct manner. Gell's central concern in the book is to show that artefacts can be conceived as possessing the kind of agency one would ordinarily associate with humans. This they do insofar as they are embedded in the 'causal nexuses' through which actors detect human intentions, whether real or imagined (1998: 33). For example, the mines buried in Cambodia are 'indexes' of Pol Pot's deadly intentions, in the sense that those unlucky enough to encounter them are caught up in a causal chain connecting the mines back to Pol Pot's violent decision to place them there (1998: 20–21). The mines themselves have 'agency', then, inasmuch as people's cognitive propensity to trace this causal chain extends Pol Pot's intention to kill onto the mines themselves, rendering them constitutive of his deadly power: the mines embody and enact the intention to kill, so in that sense they are agents. Or, to take one of Gell's less harrowing examples, a car that 'refuses' to start also has agency inasmuch as it is imagined to be obstinate when we are in a hurry (1998: 22). While there may in fact be no ghost or other form of intentionality in, or behind, the machine, the very fact that we can imagine the situation in this way allows the car to have a power over us, and hence in that sense the car has agency too.

To be sure, the idea that things can be understood as possessing agency in the same sense as humans do, for which Gell's book is so widely cited, sounds a lot like a posthumanist reshuffle of the distinction between humans and non-humans. Nevertheless, there is some ambiguity as to how far agency really attaches to things themselves in Gell's model. Indeed, in reading the book, one is never quite sure how seriously Gell wants us to take the, after all, rather scandalous notion that things can be ascribed with intentions (Pinney & Thomas 2001; Layton 2003; see also Knappett & Malafouris 2008; Bille 2013). Thus, agency for Gell is only ever an indirect attribute of things, its origins lying ultimately with a *human* agent, whether real or imagined, whose intention the thing in question only indexes – hence, for example, the significant distinction Gell makes between the 'secondary' agency of (non-human) indices and the 'primary' agency of the (human) intentions they are deemed causally to index (Leach 2007; cf. Gell 1998: 17–21). Things, for Gell, cannot *really* be agents, if by that we mean anything more than the kind of attribution of agency involved in swearing at a car for making us late. As Miller puts it in his own critique of *Art and Agency*, 'Gell's is a theory of natural anthropomorphism, where our primary reference point is to people and their intentionality behind the world of artefacts' (2005: 13). But the irony is that, precisely by virtue of this anthropomorphism, Gell's attempt to raise the status of things by conferring them with agency turns out to be more similar to Miller's than may at first appear. Where Miller enhances the role of things by making them operative in the making of human beings, Gell does so by making them operative in acts of human agency.

So, in sum: humanist approaches, which leave the ontological distinction between things and people unmodified, cannot but raise the status of things with reference to their associations with humans. The whole point about the common-sense distinction between people and things is that the former are endowed with all the marks of ontological dignity (agency, freedom, reason and so on), whereas the latter are not. So if you want to take the thing seriously in your analysis while leaving intact the ontological divide that separates it from people, then all you can do is find ways to associate it more intimately with people, thus endowing it with some of the 'dignity' and 'respect' otherwise associated only with humans. Posthumanist approaches, by contrast, posit a different ontology of people and things, thereby redefining the most essential properties of things (i.e. rather than merely redistributing properties such as 'agency' or 'intentionality' across the person/thing divide). Posthumanism, therefore, raises the hope of taking the thing seriously 'as such', although one immediately has to add the proviso that 'the thing', following its ontological reconstitution, is no longer the thing as we ordinarily know it.

One of the best-known examples of this approach is Actor Network Theory (e.g. Latour 1993, 2005). As we discussed in Chapter 1, for Latour all the entities that the 'modern constitution' distinguishes as either people or things need to be conceptually refashioned into 'hybrids': network-like knots, chains and assemblages of mutually transformative relations.

Each composing element of these relations (itself a relation, following Tarde 2012 and in line with Strathern's conception) is an 'actant' inasmuch as it has a transformative effect on the contingent and analytically localized aspect or moment of the actor network. So here agency is not the effectuation of a human intention, as it was for Gell. Rather, it is a property of networks of relationships (hybrid ones, involving all the elements that a modernist ontology would want systematically to distinguish from one another) that emerges as and when the differential and inherently relational elements that they involve 'make a difference' to each other (Latour 1988). The classic and much cited example is Latour's discussion of the gun debate in the United States (e.g. 1999: 180). Here the responsible agents, Latour suggests, are neither the guns themselves (as the anti-gun campaigners argue) nor the people who use them (as the gun-lobby would have it – 'guns don't kill, people do'). Responsibility rather must be ascribed to the hybrid assemblage, or as Latour calls it 'collective', which gun users and guns form together: the 'person-with-gun'.

There can be no doubt that, thus ontologically revised or redefined, things are indeed taken seriously. The new kind of analytical entity that Latour proposes, the hybrid assemblage of humans and non-humans in mutual transformation, is an agent in as serious a sense as one might wish to take that term: its very constitution is defined by its ability to act as such, even if at the cost of humanist questions of intentionality and freedom, which are bracketed and sidelined in the process (Laidlaw 2013: 183–6). Indeed, the bold political philosophy that Latour has built on the back of his move to collectives of things-and-people is testimony to this: 'political representation of nonhumans seems not only plausible now but necessary, when the notion would have seemed ludicrous or indecent not long ago', he writes, and raises the prospect of a 'parliament of things' (1999: 198). Evidently, worrying about the 'emancipation' of non-humans is for Latour no rhetorical hyperbole: it is as vital as fighting for the emancipation of humans ever was, in view of the impending ecological catastrophes of what he too calls the anthropocene (Latour 2014b).

Yet, Latour's quest is not free from ironies of its own. For in order to avoid emancipating things anthropomorphically, by way of their association with humans as per Miller or Gell, Latour seems to end up defining them, in what looks like a revisionist move, *as* associations (assemblages, collectives, networks). Imaginative and radical as this move is, from the perspective of the present argument its begs a critical question: how far, if at all, does the dignity, respect and agency conferred on the actants of a posthuman network or collective rub off on the things that a humanist metaphysic would call 'things'? Does Latour's revision of the ontological constituents of the world get us any closer to answering the question of how far the status of things themselves, 'as such', might be raised, allowing them to make a difference 'in themselves' to the ways in which we conceptualize them?

Of course, from a Latourian point of view, these questions are either meaningless or foolish. There is no 'thing in itself', other than in the modern chimera. To raise the very question of things' capacity to make a difference 'as such' is to engage in an act of modern purification. Nevertheless, something important tends to get if not lost, then at least muted, in the Latourian translation of things into collectives, and the ontological censorship that it entails. Namely, we would argue, the very qualities that seem most peculiar to 'things' as one ordinarily conceives of them, and in particular the aspects of things we would ordinarily tag is their material qualities, such as those studied by material scientists. Indeed, one can object that in principle Latour's basic ontological premise, namely the 'symmetry' of treating the entities that a modernist metaphysics purifies as persons *or* things as hybrid relations of persons *and* things, renders any interest in the most 'thing'-like aspects of artefacts harder to pursue.[2] Qualities one would ordinarily call 'material' are in principle in deep ontological entanglement with the (also) human

---

[2] One might here add that much the same holds for what could be seen as the most human-like aspect of humans, such as strong emotions, intense desires and other affects – not exactly topics that seem to have inspired, or lend themselves to, STS-inspired analysis (though see Suzuki 2015).

projects that they help constitute, so one wonders whether in practice a Latourian take on things allows space to disentangle them and allow them to be explored in their own right. With the metaphysical bathwater of 'materiality', it seems, goes also the baby of 'materials' as a legitimate analytical concern.[3]

This way of putting matters shows how close this concern comes to one expressed recently by Tim Ingold (2007). Impatient with what he sees as perversely abstract and intractably abstruse debates about 'materiality' in recent years, Ingold urges anthropologists to 'take a step back, from the materiality of objects to the properties of materials..., a tangled web of meandering complexity, in which – among myriad other things – oaken wasp galls get caught up with old iron, acacia sap, goose feathers and calf-skins, and the residue from heated limestone mixes with emissions from pigs, cattle, hens and bears' (2007: 9). Illustrating what in Chapter 1 we called the 'alternative ontology' approach in recent writings on ontology, Ingold makes no secret of the fact that his manifesto for a renewed focus on materials is itself metaphysically motivated, and bound up with a particular way of viewing the relationship between humans and things. Inspired by Gibson as well as phenomenology, as we

---

[3] Granted, it would be inaccurate to say that these elements do not play an important role in Latour's often highly sophisticated empirical analyses, as well as those of his followers. For example, Latour's refutation of the technological determinism of saying that 'guns kill people' does not stop him from emphasizing the particular forms of agency that a gun's technical characteristics – the mechanics of detonation, velocity, accuracy and so on – contribute to the man-with-a-gun assemblage. Nevertheless, the net effect of Latour's ontological amalgamation of such characteristics with the people they act to transform tends to render them as corollaries of projects and concerns that a lay non-Latourian account would interpret as distinctly human. To be sure, this tendency may be a contingent characteristic of the particular questions on which Latourian analyses have been put to work, rather than a direct consequence of the conceptual premises of Latour's model. Indeed, in his detailed exposition of Latour's argument, philosopher Graham Harman (a main exponent of the 'speculative realist' approaches we examined in Chapter 1) raises the prospect of a Latourian network comprising only what he calls 'object-object relations', such that, for example, one could study the manner in which 'fire analyses cotton' during combustion (2011).

saw, Ingold sees humans and things as submerged on an equal ontological footing in 'an ocean of materials' (2007: 7). He writes:

Once we acknowledge our immersion, what this ocean reveals to us is ... a flux in which materials of the most diverse kinds – through processes of admixture and distillation, of coagulation and dispersal, and of evaporation and precipitation – undergo continual generation and transformation. The forms of things, far from having been imposed from without upon an inert substrate, arise and are borne along – as indeed we are too – within this current of materials. (ibid.)

Whatever one might think of Ingold's wider theoretical and metaphysical project, this plea for materials is a powerful reminder of a whole terrain of investigation that any attempt to take things seriously cannot afford to ignore. Indeed, Ingold's plea for materials takes us to the heart of this chapter's central question, of whether it might be possible to enhance the status of things in the conceptual economy of anthropological analysis. The problem is one of wanting to have one's cake and eat it. Eating the cake, in this case, is taking fully on board the posthumanist (e.g. Latourian) point that any thoroughgoing attempt to elevate the status of things in our analyses must eschew as a starting point any principled distinction between things and humans. Having the cake is finding a way nevertheless to credit the Ingoldian intuition that a full-hog engagement with things or 'materials' must place those characteristics that are most thing-like or 'thingy' (the designation is purely heuristic, with no metaphysical prejudice, as we shall explain more later in this chapter) at the top of its agenda. To articulate a way to square this circle, it is useful to return with a critical eye to *Thinking Through Things*. For ease of reference, we shall refer to this book's argument, particularly as it was put forward in its Introduction, as 'TTT'.

Thinking Through Things

Plotted onto the trajectory of increasingly radical attempts to erase the human/thing divide, TTT could be placed at the far posthumanist extreme. Indeed, to compound an already ugly term, the argument of TTT could be designated as 'post-posthumanist', in that it takes on board the Latourian suggestion that the distinction between people and things is ontologically arbitrary, but adds (contra Latour among others) that, this being so, the solution for elevating the analytical status of the thing must not be to bind it to an alternative ontological order (e.g. that of collectives, assemblages or the Actor Network), but rather to free it from any ontological determination whatsoever. In this way, TTT brings to bear on the anthropological debate about materiality the central agenda of the ontological turn as we have been articulating it thus far in this book. We start by presenting TTT's central argument, in order then to go on critically to assess where it leaves the status of 'things' in the economy of anthropological analysis.

TTT's argument involved two key claims – one critical and one positive. The critical move, which took off directly from the work of Strathern, went as follows. If in any given ethnographic instance things may be considered, somehow, also as non-things (e.g. an artefact that contains a spirit, as per Maussian gifts, or a salt lick on a river bank that is a human dwelling, as in Amerindian perspectivism), then the notion of a thing, anthropologically speaking, can at most have a 'heuristic', rather than an 'analytical', role (Henare et al. 2007: 5–7). So attempts to analyse the things we call objects, artefacts, substances or materials in terms of their objectivity, substantiality or, as has become most popular, their 'materiality', are locked in a kind of ethnographic prejudice. This is because they skew the analysis of things with conceptualizations that, from an ethnographic point of view, may well be entirely alien to them. How could one decide in advance, *before* engaging with it ethnographically, that, say, the cloak worn by a Mongolian shaman is best conceived as a 'material object' (Pedersen 2007)? Indeed, the same reservations hold also for attempts theoretically to 'emancipate' things by attributing them a priori with all sorts of qualities earlier analytical approaches would take to belong only to humans, such as sociality, spirituality, intentionality and again, most popularly, agency. Such attributions cannot be made by theoretical fiat. Whether or not they may be appropriate, and in what sense, is first and foremost an ethnographic question.

If what things are is to be treated as an ethnographic variable in this way, then the initial analytical move must not be to 'add' to that term's theoretical purchase by proposing new ways to think of it – e.g. things as sites of human beings' objectification (Miller), as indexes of agency (Gell), as momentary assemblages of heterogeneous hybrids (Latour), or what have you. Rather, the strategy must be one that is capable of effectively *de-theorizing* the thing, by emptying it out of its many analytical connotations, rendering it a purely ethnographic 'form' ready to be filled out contingently, according only to its own ethnographic exigencies. This, then, is what treating the thing as a 'heuristic' means: using ordinary, unreconstructed, 'modern' ontological assumptions just as a *tag for identifying it as an object of ethnographic study.* This, according to TTT, is the prime step towards allowing things to dictate their own terms of analytical engagement: to be able to talk about them 'as such', without allowing the language we use to prejudice our analysis of what they might actually *be.* It is very much in line, then, with the central strategy and objectives of the ontological turn.

If according to the TTT argument half of the way towards raising the analytical status of things is to empty out the notion of 'thing' of its contingently a priori metaphysical contents (Step 1: 'thing-as-heuristic'), then the other half is to formulate a way of allowing it to be filled by (potentially) alternative ones in each ethnographic instance. This was the second and positive move of the TTT argument (its Step 2), which was captured by a methodological injunction: 'concepts = things'. This second move can be described as complementary in that it follows directly from the issue that motivates the heuristic approach in the first place, namely the possibility – and in so many instances the fact – that the things we anthropologists[4] call 'things' might not ethnographically speaking be things at all, or not in the way might initially assume them to be.

For note that in TTT the 'concepts = things' injunction was determinedly *not* proposed as some new theory of the thing (or of concepts for that matter, *pace* Viveiros de Castro's search for an anthropological concept of the concept – e.g. 2003). The idea was emphatically *not* to provide some kind of revisionary metaphysic or alternative ontology, to the effect that, where people have so often assumed things and concepts to belong to opposite ontological camps, we should from now on recognize them as belonging to the same one (viz. the kind of approach Latour and Ingold may be said to advance in different ways, and which some critics have read also onto TTT – e.g. Graeber 2015; Pickering 2016; Domínguez-Rubio 2016). To the contrary, TTT's 'concepts = things' formula was offered as a methodological clause for sidestepping just such theoretical prescriptions. In particular, it was supposed to foreclose a very real danger in the ethnographic study of things, namely that of imagining them as different ways in which people may think 'about' them (represent them, imagine them, socially construct them and so forth). For to think of things in that manner is just a way of endorsing the basic 'modern' tenet of one-nature-many-cultures in what may be its crassest version, namely the idea of inert and mute things invested with varied meanings only by human fiats of representation. It is, in other words, to rule out of court the very possibility in which we are interested here, namely that things themselves might be able to help to provide *alternative* ways for us, as

[4] Note that the pronoun 'we' here should under no circumstances be understood as a cultural, social, historical or geopolitical designation. In line with the overall argument of this book, which holds the ontological turn as a strictly methodological proposal concerning the reflexive relationship between objects and terms of anthropological description, the word 'we' refers strictly to the position of the anthropologist as analyst dealing with ethnographic materials. While hardly redundant for other purposes no doubt, questions regarding where the 'we' might come from, whether it is male or female, Western or not, and so on, miss the point we (the authors now!) are making.

analysts, to conceptualize what they are – ways that challenge and go beyond our metaphysical (and by that token also methodological and analytical) expectations.

So the 'concepts = things' clause is meant to placate the dangers of metaphysical prejudice, by turning the ontological turn's central concern with conceptual definition – the concern with transforming concepts – into a methodological formula. Simply: instead of treating all the things that your informants say of, do to and with things as modes of 'representing' the things in question, *treat them as modes of defining what those things are.* Once again, it is of critical importance to emphasize here the methodological nature of this injunction. Treating what informants say and do around things as manners of defining what those things are is not meant as a theoretical statement about how things are constituted or 'enacted' as 'active entities' or 'actors', as Law and Mol among others have argued (2008; cf. Bille 2015). It is meant strictly as a methodological injunction regarding the way things and what informants say and do around them should be treated in the context of their anthropological analysis.

The immediate advantage of this way of proceeding is that it renders wide open precisely the kinds of questions that lie at the heart of the concern with raising things' analytical status, and indeed the ontological turn more generally, namely questions about what kinds of things 'things' might be in any given ethnographic case. If every instance anthropologists would deem a different 'representation' of a thing is conceived as a potentially different way of defining what such a thing might be, then all the metaphysical questions about its character *qua* 'thing', what 'materiality' might be, 'objectification', 'agency' – all that is now up for grabs, as a matter of ethnographic contingency and the analytical work it forces upon us. Instead of merely offering sundry ways of confirming the base metaphysic of mute things invested with varied meanings by humans, the concepts-as-things approach holds up that very ethnographic variety as a promise of so many ways of arriving at alternative metaphysical positions – *whatever* they might be.

Rethinking Things

TTT takes us a long way towards the kind of engagement with 'the thing' that we are looking for in this chapter. Indeed, it may even seem that, taken together, its two central methodological steps – 'thing-as-heuristic' and 'concept = thing' – take us all the way there, effectively opening up the space for things themselves, as one encounters them in any given ethnographic situation, to make a contribution in their own right to the way in which we conceive them analytically – to help dictate, in other words, their own terms of analytical engagement. Nevertheless, such a conclusion would be too hasty. For where exactly, one may ask, does this argument leave us with respect to the 'thingness' of things? How far does the methodological argument of TTT make a virtue of those aspects that are most characteristic of things as we ordinarily think of them, namely their material properties?

It is telling that the sometimes flamboyantly programmatic pronouncements of TTT's Introduction made little mention of the material properties of things. At any rate, the way material attributes featured in the ethnographic accounts of people's own ways of 'thinking through things' in the chapters that followed was left largely unspecified in the Introduction's otherwise overtly methodological argument. Instead, bearing out the volume's subtitle ('theorizing artefacts ethnographically'), all the emphasis was on *the ethnography* of things, and particularly the ethnography of *the people* to whom things matter in such diverse ways. So we may ask: If what motivated TTT's approach is the fact that in varied instances people speak of or act with things in ways that contradict our assumptions about what a thing might be; and if, furthermore, it is just those ways of speaking about, and acting around, things that are supposed to provide the 'content' of their potentially alternative metaphysics; then in what sense, if at all, do the things *them-selves* make a difference to the manner in which they are conceptual-ized by the anthropologist? It would seem that the leverage for thinking out of the metaphysical box that TTT sought to provide was owed, to a large extent, to the chapters' ethnographic magic, to coin a phrase, rather than the specifically 'thing-like' character of their subject matter (see also Alberti 2014b).[5]

This orientation towards the ethnography of things, as opposed to the things themselves, flows directly from TTT's alignment with the central arguments of the ontological turn, and particularly the eman-cipatory political agenda about taking people seriously that Viveiros de Castro sought to pursue in developing them. Indeed, if for Viveiros de Castro the emancipation of peoples in anthropology is a matter of opening up space for their 'conceptual self-determination', then TTT's proposal amounted essentially to the addendum that the ethnography of people's engagement with things is a prime site for pursuing such a goal (Henare et al. 2007: 8–12). In other words, whatever enhancement of things' analytical status TTT might offer was rhetorically subsumed under the political agenda of conceptually emancipating people. TTT's two-step methodology reflects this directly. The 'thing-as-heuristic'

---

5 It emerges, then, that TTT's claim to raise the status of the thing along the lines discussed previously is open to a critique that is analogous to the one advanced earlier in relation to Latour. Latour, we saw, emancipates the thing by entangling it ontologically with persons – subsuming both under the terms of his revisionary ontology of networks comprising people-and-things. TTT does something similar, though now at the level of analytical methodology. It raises the status of the thing by entangling it heuristically with all that the people concerned with it say and do around it, subsuming things and their ethnographic accounts under the terms of a revisionary methodology. Indeed, just as a Latourian might object that to demand an emancipation of the thing 'as such' is flatly to deny the significance of Latour's ontology of collectives, so one might want to contend that such a demand similarly contradicts TTT's methodological injunction of 'concept = thing.' As far as TTT is concerned, things as such just *are* what our ethnographic descriptions of them define them to be. But our question here is precisely whether things, *heuristically defined*, might be able to contribute to their own conceptual variation.

move opened up 'things' as a locus of ontological self-determination, while the 'concept = thing' clause allowed the *ethnography* of what people do and say around their things in different parts of the world to provide these things with ontologically variable contents. Emancipation by ethnographic association, as it were.

Yet, the Ingoldian challenge remains: what of materials and their properties? This brings us to the crux of our argument in this chapter, which is that, suitably reconsidered and extended, the line of thinking that TTT initiated *can* make full analytical virtue of things' material characteristics. If TTT tended to bind together the question of what difference things, qua things, can make to the terms in which their anthropological analysis is conducted with the question of how those things matter to *people* in any given ethnographic setting, our task here is heuristically to disentangle things from people, in order to explore how far and in what ways the former make their own kinds of difference to the way they can be conceptualized. In this sense, our task is to isolate and clarify a methodological potential that was already implicit in TTT, though in a submerged and unclear way.

The confusion, we suggest, lies in the symmetry of TTT's second methodological move, namely the formula 'concept = thing'. For the formula can be read in two directions: if concepts can define things, then things can also define concepts ('concepts = things' $\Leftrightarrow$ 'things = concepts'). In TTT the second reading was assimilated to the first, and thus remained submerged. Bringing it to light, we suggest, puts at the heart of the matter the question of things' capacity to generate concepts, rendering it an immediate corollary of TTT's methodology. Indeed, if in TTT the formula 'concept = thing' expressed the possibility of treating what people say and do around things as manners of defining what those things are, then the reverse rendition of this formula, 'thing = concept', raises the prospect of treating that thing itself as a manner of defining what we (analysts now, rather than natives) are able to say and do around it.⁶ At issue, to coin a term, are the thing's 'conceptual affordances' (cf. Dant 2005: 70–82).

To see the significance of this it pays to consider a little further the consequences of reversing the 'concept = thing' formula to give 'thing = concept'. As we have explained, with its two-step methodology, TTT grounded conceptual experimentations in ethnographic contingency. Having emptied the notion of 'the thing' of any conceptual presuppositions about what may count as a thing in the first place, we then fill it back up with alternative conceptualizations drawn from the ethnographic data we find around it, which in turn provide the reflexive empirical source for subsequent acts of anthropological

---

⁶ The present argument about the conceptual affordances of things, as we shall be calling them, is meant as a contribution to the reflexive line of thinking we detect in (and as) the development of the ontological turn, which concerns the ways in which the ethnographic materials anthropologists study (in this case 'things') can have an effect on the terms in which they are studied. In *TTT* this agenda was sometimes run together with what we would now treat as a separate (though related) one, namely the much better-explored question of how material artefacts influence the ways in which the people that engage with them think about them, or conduct conceptual operations in general. We have in mind here the longstanding and inter-disciplinary literature on the material conditions of possibility of mental processes, centring on such topics as material, extended or distributed cognition (e.g. Hutchins 1995; Clark & Charmers 1998; Clark 2008; Menary 2010; Malafouris 2013), the relationship between thought, skill and environment (e.g. Gibson 1979; Lemonnier 1992; Ingold 1997, Conneller 2011), the materiality of memory and the imagination (e.g. Munn 1986; Küchler 2002; Sneath et al. 2009; Bille, Hastrup & Flohr 2011; Chumley and Harkness 2013), and more. Undoubtedly, the argument that follows here about how things can contribute to their own analysis by virtue of their (heuristically identified) material properties could be brought to bear on the broader anthropological conversation about the role of things in conceptual processes more generally, and we shall be making some remarks in this regard in our case-studies that follow. Note, however, that, in line with our point about the ontological turn's figure/ground reversal between 'anthropology' and 'the world' in the previous chapter, transposing our argument in this way onto a broader concern with how things relate to people's concepts would have to first involve a contingently reflexive interrogation of what 'things', 'people' and 'concepts' (all now deemed as flesh-and-blood constituents of the world, rather than heuristic placeholders for the constituents of anthropological analysis) might amount to in any given ethnographic situation.

conceptualization. But now we may ask: what would be the equivalently empirical grounding of the reverse procedure that the 'thing = concept' clause seeks to articulate? Following through on the symmetry of the reversal, we suggest that the answer must be found in the material characteristics of the thing itself. In TTT, the 'concept = thing' principle grounded anthropological conceptualizations empirically by sourcing them in the ethnography of what people say and do around them. That is to say, people's so-called beliefs and practices were treated, methodologically, as manners of *defining* what the things in question are. So the present reversal involves a symmetrical, but inverted, relationship between empirical source and conceptual outcome. Now the 'thing' (tagged heuristically, following TTT, as a 'material object') becomes the empirical source of conceptualization – and in this case, since what the thing itself *might be* is what is at issue, also of its *own* conceptualization. With what other 'stuff' can things feed their conceptualizations, after all, than the very stuff that makes them what they are, as 'things'? So, the data that make a (conceptual) difference, in this case, are no longer what people say and do around things, but rather what we hear, see, smell, taste and touch of the thing as we find it (heuristically) as such. Just those aspects of material objects, in other words, Ingold challenges anthropologists to deal with.

The difference from Ingold, however, is that his interest is in celebrating this material and sensuous level of things in its own right, to explore things' mutual 'enmeshment' with people and other organisms, as well as their 'affordances' (following von Uexküll) for living beings in the broader ecology of life, as we saw in Chapter 1. By contrast, in raising the question of the *conceptual* affordances of materials and their properties, our interest is not in the ecology of their material alterations but rather in the economy of their conceptual transformations as part of the process of their anthropological analysis: how their material characteristics can help to form their anthropological conceptualization. At issue is not the enmeshment of materials in forms of life, but rather materials' transformation into forms of analytical thought.

Such a notion of transformation transposes onto the study of things the postplural mode of abstraction that lies at the heart of Strathern's anthropological project. As we saw in Chapter 3, for Strathern, anthropological comparison involves a form of abstraction that renders the potential for conceptualization *internal* to the ethnographic materials under investigation. Strathern, we showed, 'scales' ethnographic materials, by releasing from within them their potential for self-transformation in the direction of what an unreconstructed 'plural' metaphysic would call 'concepts'. So if in Chapter 3 Strathern's scalings were shown to be what abstraction looks like when it is divorced from the ontological distinction between concrete (ethnographic materials) and abstract (concepts), then the 'thing = concept' clause we are proposing in the present chapter transposes this Strathernian idea (or scales it, if you like) in the direction of things. Where the humanist things/concepts binary posits abstraction as the ability of a given concept to comprehend a particular thing, external to itself, in its extension, the posthumanist formula 'thing = concept' casts this as a movement internal to the thing itself: the thing differentiates *itself*, no longer as an instantiation 'of' a concept, but a self-transformation *as* a concept. It invents itself (in Wagner's terms) by transforming its own intension (to use Viveiros de Castro's).

In what remains of this chapter we illustrate how such an analysis might proceed, with reference to two concrete bodies of ethnographic material drawn from our own respective chapters in TTT, namely Holbraad's account of the role of powder in the Afro-Cuban divinatory tradition of Ifá and Pedersen's ethnography of the role of cloaks in the practice of shamanism in Mongolia. We begin with Holbraad, showing how his original analysis exemplified TTT's two-step formula as it stood, proposing a conceptualization that took seriously what Cuban diviners say and do with *aché*, the prime ritual ingredient and cosmological principle of Ifá divination. We then go on to explore ways in which a concern with material properties also featured in this analysis, albeit in a submerged form. While Holbraad did not clarify the methodological implications, here we show that the material properties of *aché* play a decisive role in his analysis. So, in the section that follows, our task will be to identify and clarify the analytical effects of material properties in the context of Holbraad's TTT argument, showing how they can be understood with reference to the third step we have added to the TTT methodology, 'thing = concept'. To ameliorate the inelegance of this lengthy self-citation, which we hope the reader will indulge for the sake of the argument, in the section that follows we paraphrase freely from Holbraad's original chapter (2007), flagging up the points at which our critical account diverges from it.

Powder and Its Conceptual Affordances

Aché is a West African, *mana*-type term that Afro-Cuban diviners use to talk in two senses[7] that initially appear quite different. On the one hand they use *aché* to refer to the power (in Spanish, *poder* or *facultad*) they have as diviners to make deities appear during Ifá divination. On the other hand, they use the term to refer to a particular kind of consecrated powder that is deemed to be a necessary ritual ingredient for achieving these divine appearances during the ceremony of divination. As far as the practitioners of divination are concerned, the terminological coincidence corresponds to a real connection. Diviners themselves account for their divinatory powers ('their *aché*', in that sense) partly with reference to their privileged access, as initiates into the secrets of divination, to the special consecrated powders they use (*acheses de Orula*, as they call them in the plural, in reference to the patron-deity of divination Orula, to whom the consecrated powders are deemed to belong). Conversely, they

---

[7] *Aché* has other, related, meanings also, which Holbraad explores in his TTT chapter (2007: 201–2).

distinguish those powders from others (e.g. those used by initiates of other Afro-Cuban religious traditions, including sorcerers who, among other secret ingredients, use their own powders to harm their victims) by pointing to the special consecration procedures they have undergone to gain the appropriate divinatory powers.

One might say, then, that the double formula for Ifá diviners is 'no powder no power' and 'no power no powder.' Their power lies in the powder while, conversely, the powder *is* power. We arrive, then, at a counter-intuitive suggestion of the order of the classic anthropological debating-line, 'twins are birds' (Evans-Pritchard 1956) – the kind of notion that, according to Viveiros de Castro, anthropologists are prone not to 'take seriously', as we saw in Chapter 4. Similarly to what we saw with gifts there (and the same argument could be made of twins and birds – see Venkatesan et al. 2010), if we know what powder is at all, we know that it is not also power in any immediately meaningful sense (it's just powder!), and much less can we accept that power might also be just powder (of all things!). Hence, here too, the classical kind of anthropological question arises: why might Cuban diviners 'believe' such an odd idea? In fact, as we also saw in our discussion of gifts, the point is stronger than that: for as long as our analysis of *aché* remains within the terms of an axiomatic distinction between things and concepts, we *cannot but* ask the question in these terms. We know that powder is just that dusty thing diviners use with their fingers. So the question then becomes why Cubans might 'think' that it is also a form of power. How do we explain it? How do we interpret it?

Of course, one could treat the distinction between concepts and things merely as a heuristic device, as per TTT's first step discussed earlier, which was what Holbraad did in his chapter. This creates the analytical space to ask questions about that powder one would intuitively identify as a 'thing', without prejudicing the question of what it might be, including questions of what it being a 'thing' might even mean. Answers to such questions, then, can be found in the ethnography of Ifá divination by examining people's apparent 'beliefs' about this powder, including the key notion that it is a form of power, by treating them rather as *definitions of what powder is*, as per the second move of the TTT method (concept = thing). So: Cuban diviners do not 'believe' that powder is power, but rather *define* it as such. Significantly, this way of setting up the problem raises the metaphysical stakes. To the extent that the default assumption is that powder is *not* to be defined as power ('it's just a dusty thing', one might want to say), the anthropological challenge must be to reconceptualize the notions of 'powder' and 'power', along with their many ethnographic and analytical corollaries (e.g. 'thing', 'concept', 'divinity') in a way that would render the ethnographically given definition of powder as power amenable to an anthropological description that makes sense of it. Very much in line with our overall argument here about the ontological turn, Holbraad's TTT chapter sought to do just that, suggesting that in order to make sense of the mutual entailment of powder and power one must think of the divinities that the powder has the power to reveal during divination as motions rather than entities, as we shall see presently.

Now if, as we argued previously, the problem with TTT was that it raised the analytical status of the thing only by associating it ethnographically with a conceptually emancipated native, then Holbraad's analysis of *aché* was very much an instance of this. We have already seen, for example, that the problem that the chapter was devoted to solving – what might a powder that is also power be? – was ethnographically as opposed to 'materially' driven: it was not powder that told Holbraad it is power, it was his diviner informants. Indeed, a host of ethnographic data was used to frame and develop the problem itself, as well as parts of its analytical solution. In particular, this involved providing an account of Afro-Cuban divinatory cosmology based on informants' responses, to show that for diviners power consists above all in the ability to render otherwise absent divinities present during the divinatory ceremony. This power, it was shown, manifests in divination as the 'signs' the diviners mark with their fingers on the powder that is spread in the surface of their divining board, which are called *oddu*, and are said to 'be' divinities in their own right. Then, on the basis of this ethnographic information, Holbraad went on to show that the notion of a powder that 'is' power emerges as a solution to a version of the age-old theological conundrum, familiar in the anthropology of religion (e.g. Keane 2007): apparently 'transcendent' deities are rendered 'immanent' on the surface of the divining board, allowing those present in the divination to relate to them directly. So the power of powder in Ifá resides in the manner in which it effectively, and very practically, *solves* what we may think of as an Afro-Cuban equivalent to the (perhaps more familiar) Christian 'problem of presence' (Engelke 2007), namely that of allowing otherwise absent divinities to become present (in this case on the surface of the divining board).

So far, then, so ethnographic: an understanding of the notion of power in Ifá, and particularly of its connection to powder, is built up by taking seriously what the practitioners have to say on the matter. Note, however, that this ethnographic account serves only to *set up* the central analytical challenge posed by Ifá practitioners' understanding of *aché*, namely that something as seemingly prosaic as powder is deemed to have the power to render the divine present. Indeed, if this reminds us of the notoriously intractable problem of divine transcendence in Christian theology, then the matter-of-fact, entirely practical way in which *aché* powder solves this for the diviners appears all the more puzzling. It is at this point, then, that the basic 'move' of the ontological turn becomes pertinent. How do we conceptualize the problem of transcendence to make sense of powder's power to solve it? What might divinities that can be rendered present in this way *be*? In what way, and in what sense, is powder able to make all this happen?

It is just at this point, we suggest, that Holbraad's analysis takes a turn *away* from the ethnography of what diviners say, zooming in instead on the powder 'itself', to focus on its material properties. Yet, in Holbraad's original analysis this move was treated as a consequence of taking the *diviners* seriously. If for them power is powder, he suggested, then to understand what this power is and how it works we should look at the powder and what it does by virtue of its material features (2007: 206). Seen from the point of view of our present argument about the conceptual affordances of things, however, what is obscured by Holbraad's original insistence on 'taking people seriously' as the prime tenet of the ontological turn, is that this requires us *also* to take seriously the thing itself, by the same token. In other words, the notion that powder has the power to reveal divinities (taking people seriously) leads to the prospect that powder may also have a different power altogether, namely the power to reveal the *concepts* we need in order to make sense of it (taking the thing seriously). In this way, the ethnography of what diviners say and do points beyond itself, as Holbraad also put it in TTT, 'to the things themselves' (2007). From the cosmological power of powder we get to the *analytical* power of powder.

So, although Holbraad did not clarify this in his chapter, his analysis required reading the 'concept = thing' formula also in the opposite direction, 'thing = concept', as formulated earlier. Indeed, while the former step served to set up the analytical conundrum, as we have seen, the latter effectively delivered its solution, inasmuch as Holbraad ended up in his original analysis extracting a conceptualization of power from the most prosaic material characteristics of the powder itself. If, taking seriously what diviners say, the power of powder resides in revealing the *oddu*-divinities on the surface of the divining board on which it is spread, then to understand how best to conceptualize this power must involve taking seriously the manner in which it is enacted as a practical, and entirely concrete and (again, speaking heuristically) material operation on the divining board. This was the argument's strategy, even though Holbraad did not clarify its methodological significance.

In particular, focusing on the material properties of the powder, Holbraad showed that the *oddu* are revealed on the powder's surface by virtue of its material *perviousness*. As a collection of myriad specks of dust – a 'pure multiplicity' (2007: 207) –, the powder is 'marked' (*marcar*) as the extensive movement of diviner's fingers on its surface produces an intensive displacement of the powder. So the powder reveals the divinities like Archimedes' bathwater measured volume: the figures of the *oddu* are 'registered' (*registrar*), as the diviners themselves say, as temporary displacements of the ground that the powder provides. From the point of view of its material properties, Holbraad concluded, the power of powder lies in its motility: its capacity to be displaced so as to reveal the *oddu*-divinities as motions, registered on its surface by way of an entirely concrete figure-ground reversal.

The *analytical* revelation that powder provides, then, is that the divinities that it has the power to reveal are themselves *motions*. While submerged within Holbraad's original argument, the operative syllogism runs like this: The divinities 'are' the marks on the divining board, as the diviners tell us (taking people seriously, concept = thing); but those marks on the divining board's powder (*qua* thing) are motions; therefore (taking the thing seriously, thing = concept), the divinities *are* motions. Spelt out in this way, the argument turns most crucially on extending the TTT formula to include the conceptual affordances of things. For it is just this materially derived concept of motility that delivers the solution to the ethnographic conundrum about powder's power to reveal divinities. If divinities are to be conceived as motions (rather than as 'entities', for example), then the age-old theological puzzle of how a god might pass from a state of transcendence to a state of immanence is resolved into a question of relative distances. Conceived as a state of motion, transcendence becomes a matter of distance and immanence a matter of proximity. The divinatory task of rendering the divinities present during the divination is no theological mystery. *Qua* motions, *oddu*-divinities *just are* the capacity to traverse distances – they are *themselves* traversals. The diviners' task, then, is to *direct* these inherently motile divinities so as to bring them 'close' enough to reveal themselves. While this process of invocation is a delicate and complex affair, involving more means of elicitation than just the powder (e.g. herbal infusions, consecrated items of other kinds, words, songs, libations and sacrificial offerings – see Holbraad 2012: 130–43), the powder is indispensable because it acts as the catalyst of that all-crucial, and final, moment of revelation, as the *oddu* appears on the divining-board, there for all to see. The power of concretion, one might say.

Somewhat despite itself, then, Holbraad's TTT argument serves to illustrate the manner in which things are able to evince their own conceptual effects in anthropological analysis. Although framed within a broader ethnographic argument about how things matter to people (how powder, in this case, matters to Afro-Cuban diviners), the most crucial elements of Holbraad's attempt to conceptualize the powder and its power – the ones that make the decisive difference to the conceptualization – stem from the material properties of the powder itself. If ethnography carries the weight of the analytical problem, we might say, it is the material properties of powder that provide the most crucial elements for its solution: its pervious quality as a pure multiplicity of unstructured particles, amenable to intensive movement, like the displacement of water, in reaction to the extensive pressure of the diviner's fingers, and so on. Each of these material qualities inheres in powder itself, and it is by virtue of this material inherence that they can engender conceptual effects, setting the parameters for the anthropological analysis that they 'afford' the argument.

As an irreducible element of the analysis of *aché*, then, it is *powder* that brings the pivotal concepts of perviousness, multiplicity, motion, direction, potential and so on into the fray of Holbraad's anthropological argument, as conceptual transformations *of itself*, as per the 'thing = concept' clause. In that sense, powder has the power to contribute to its own analysis – to analyse itself. And, as we are about to see, a similar conclusion – that certain artefacts are imbued with conceptual affordances that make them 'auto-analytical' – can be reached with reference to Pedersen's material on Mongolian shamanism.

Talismans as Thought

In Northern Mongolia, many people and households are in possession of talismans of shamanic spirits.[8] As elsewhere in the Mongolian cultural zone, these artefacts are called *ongod* (or sometimes *ongon shüten*). Barring possibly the notion of shaman itself (*böö*), *ongon* is the most important and most complex concept in Mongolian shamanism. Much like the notion of *aché* in Cuban Ifá, *ongod* in Mongolia refers both to shamanic spirits in their invisible, 'transcendent' form and to their multiple visible manifestations as talismans and other material forms. Indeed, shamanic spirits may take almost any shape, ranging from derelict ruins of communist infrastructure to the flow of gossip in a community (Pedersen 2011: 201–4). But their most common materialization is in the form of 'owners' (*ezed*) of mountains and other sacred places (including the burial sites of deceased shamans), in the bodies of wild animals (bears, foxes, deer etc.), and, not least, in the form of talismans.

A typical talisman consists of multi-coloured cotton strips, ceremonial silk scarves (*hadag*), leather strings, odd pieces from tools, weapons or similar metal artefacts, as well as bits and pieces of fur, teeth, bones, claws and beaks from different wild animals; all pieced together to form a complex 'bundle' of diverse materials. Some talismans are kept inside the home, as in the case of 'lineage talismans' and 'household talismans'; others, such as 'hunting talismans', are kept outside. People interact with their talismans in similar ways and mostly for the same reasons: they pray to and present offerings to them when noteworthy events happen or are about to happen in their lives (e.g. if someone falls ill, or if a member

---

[8] What follows draws on Pedersen's TTT chapter (2007) and recent writings on Mongolian shamanism (2011, 2014).

of the household is about to depart on a hunt or a long journey). All talismans should be designed or consecrated by shamans, and may be commissioned both by households and by individuals (in either case, they will eventually be passed down through the generations via either male or female lines).

The talismans kept by shamans are larger and even more complex than those kept by ordinary individuals and households. But shamans are also in possession of another artefact imbued with still greater complexity, namely the shamanic costume, comprising boots, gown, headgear, drum and drumstick.[9] Also known as the shaman's 'armour', this costume protects the possessed shaman by absorbing the souls (*süns*) of spirits into its many layers, so that they do not penetrate his or her body too deeply. Yet, at the same time, it exposes the shaman to the lethal risk of becoming lost in the world of the sprits. Unlike the ordinary nomad's gown (*deel*), which encapsulates its wearer in a protective enclosure with a minimum of openings, the shamanic gown – which is not worn with the otherwise ubiquitous sash, and from whose exterior cotton knots, strings, and flaps point antenna-like in all directions – invites maximum engagement with and intervention from the surroundings. Indeed, the gown is a whole world, or several worlds, in its own right. It thus comprises materials, textures and substances that invoke a multitude of realms and dimensions, including the forest and its different wild

---

[9] As Joe Ellis puts it, the 'shamanic gown is an extremely complex and heterogeneous piece of clothing [...where] each single element, every addition to [it], *every stitch*, h[olds] significant meaning' (2015: 58–9). So, Ellis goes on to suggest in an argument that draws on our work, 'the nature of the shamanic gown ... directly affords a theoretical account of the shamanic gown' (2015: 57). Yet, he contends, to fully harness this potential 'of allowing objects to "speak"' (2015: 57), it is necessary to take issue with what he identifies as two problems with the ontological turn, namely its 'extremely synchronic' account of 'the power of objects' and the fact that it assumes an 'underlying ontological regime' (2015: 56; c.f. Bille 2015). In the re-analysis of Mongolian shamanic artefacts that follows, we shall see how engaging with these two objections will allow us to better understand what 'analysis' or 'thinking' becomes once it is experimentally reimagined as a property of certain things.

animals, the ninety-nine shamanic skies (*tenger*), and, crucially, past spirit possessions. During every curing (*zasal*), clients or their relatives thus add ceremonial scarves and streamers (*mogoi*) in the form of knots tied into several strip-rolled cotton tassels (*manjig*) which are fastened and attached to the inside of the shamanic gown. The effect, as Caroline Humphrey has remarked, is that the performing shaman is rendered a 'knot of knots' (1996: 270) whose gown offers a sort of material testimony or 'map' of the totality of misfortunes that have prompted clients to be cured (viz. to tie knots) by the shaman in question over time.

Thus far, our account of Mongolian shamanic talismans has been fully consistent with Pedersen's TTT contribution. In the remaining part of the section, however, we reanalyse these artefacts in a way that both calls to mind but also differs in focus from the above re-analysis of powder in Ifá divination. Motivated by a desire to extend Pedersen's original analysis of Mongolian shamanic artefacts further towards the 'thing = concept' strategy discussed previously, in what follows the ultimate ambition is to explore what 'analysis' *itself* might be once it is reimagined as a capacity of shamanic talismans, rather than just of the anthropologists who study them or, for that matter, the shamans who wear them. However, to do so we first need to unpack two levels of analysis which were rolled together in Pedersen's TTT chapter (see also Gad, Jensen & Winthereik 2015). For while Pedersen's original objective was also to pose ontological questions as an analytical strategy, he was not as explicitly concerned as the Introduction and Holbraad's chapter in TTT with the effects that such a heuristic could have on the infrastructure of anthropological thought. Rather, his focus was more on the cognitively augmenting role shamanic artefacts played for the shamans and their clients – and only secondarily, and implicitly, on their purchase as anthropological concepts in their own right. In what follows, then, we begin by revisiting Pedersen's ethnography and argument to decouple the ways in which shamanic artefacts serve as 'talismans of thought' (as his TTT chapter title would have it) for local people and for anthropologists, respectively. Having done that, we can then turn to the overarching question of whether and how these artefacts may be said to constitute forms of analysis in themselves – talismans, then, *as* thought. Indeed, as we are going to see, to ask what talismans 'theorising themselves' might mean will involve probing deeper into the material properties of these artefacts, putting the 'thing = concept' formula to new use in this context.

We may start by explaining how the shamanic gown provides people attending the curing ritual with a kind of 'map' of the local distribution of misfortune, and how in that respect it serves as a 'talisman of thought' for members of the community. Recalling that each misfortune cured by the shaman is 'knotted' onto his or her gown during the ritual, this artefact gives people in the audience momentary access to knowledge otherwise hidden to them. Through a temporary reversal of what is visible and invisible, inner and outer, the performing shaman exposes the otherwise hidden intentions (*sanaa*) or inner 'layers' (*davhar*) of fellow human beings and bodies. While the victims of witchcraft and malicious gossip (*hel am*) are always present at rituals as clients afflicted with harm (*horlol*), the local perpetrators of the witchcraft may well not be. Nevertheless, the effect of their dark intentions (*har sanaa*) is rendered visible in the new knots that are on each such occasion fastened onto the shaman's gown.

In that sense, the performing shaman is an ordinary person turned inside out. By literally wearing his or her clients' misfortunes on the skin as he dons the gown, the shaman reveals what cannot otherwise be gauged from a person's appearance, namely a hidden propensity for greed, envy and violence (interestingly, the point of possession is marked by shamans making vomiting sounds, as if their insides were turned inside out as the *ongod* enter). Indeed, it is said that shamans 'have two bodies' – an ordinary human body and extraordinary, shamanic body (*böö biye*). Small wonder, then, that people tend to take such a close interest in memorizing all recent changes made to shamans' gowns during séances, for this provides them, due to the detailed symbolism of the aforementioned knotting practice, with a sort of 'database' of the distribution of misfortune in the community.

But there is another and more general cosmological sense in which these artefacts provide local people with vital shamanic knowledge. In order to show this we need to take a closer look at the shamanic cosmos, and particularly the role of multiplicity as the defining characteristic of the spirits it contains. *Ongod* exist simultaneously as 'owners' of places in the landscape or wild animals, as ancestral souls that enter the possessed shaman's body and mind, and as transcendent 'skies' (*tenger*). But most importantly, *ongod* are singularities endowed with unique spirit biographies and personalities, which derive from (but are not identical to) the dead shaman whose souls they have absorbed as guardian spirits. Shamans master a limited number of such guardians that may manifest in any physical form or bodily shape. These manifestations are known by a range of designations, such as 'metamorphosis (*huvilgaan*), 'escort' (*daguul*), 'helper' (*zarch*), 'light body' (*höngön biye*) and 'path' (*güyeel*). In this sense a given spirit comprises, or more precisely 'adds up to', an ever increasing and potentially infinite number of metamorphoses: 'The multiplicity of *ongon* rests upon the adjunction of the ancestral shamanic *ongon*'s auxiliary spirits, [that is] the different forms it can adopt, which in shamanic terminology are often known as "servants" or "metamorphoses"' (Even 1988–89: 115). A given *ongon*, then, is a manifold that is irreducible to a singular form, which moves along an unpredictable 'path'. This allows, or even compels, each spirit guardian to absorb ever more forms in its journey from one metamorphosis to the next.

These characteristics help to account for the seemingly paradoxical relationship between the *ongod*, as purely ephemeral beings, and their material objectifications in the form of talismans, which, as pointed out earlier, are also known as *ongod*. Instead of thinking of shamanic sprits as singular and stable entities, one should think of them as capricious assemblages that are irreducibly heterogeneous and multiple. In that sense, these occult entities need new elements to be constantly added to the effigies of them that people keep and attend to, for otherwise the *ongod* would simply not be able to continue being what they are supposed to be, namely phenomena that are defined by an ever increasing degree of complexity and multiplicity. At the same time, and crucially, peoples' offerings to their talismans, as well as the knots that clients add to shamans' gowns during séances, may be understood as attempts to depict, within the visible dimension of material things, the occult movement of the shamanic spirits across the invisible realm of the so-called skies (*tenger*).

Talismans thus emerge as simultaneously material residues of and immanent preconditions for the shamanic spirits' transcendent capacity to move, not unlike the hazy trace left behind from atomic particles shot through a gas chamber. Indeed, there is a sense in which the spirits *are* change in a particular register, more than they can be described as discrete entities that are imbued with the propensity *to* change in a certain way, in a sense that comes close to the one outlined earlier in relation to the divinities of Ifá in Cuba (see also Holbraad 2012). Accordingly, it could be argued that the only way to 'see' the spirits is through their absence, for their peculiar mutable 'essence' can only be gauged from the gap, or interval, between any two material iterations. Thus spirits may be said to be distributed *across* (as opposed to *in*) time. Because of their inherent capacity to differ from themselves (and not just differ from other things), time is, so to speak, hardwired into the *ongod* – temporality and transformation is internal to their very form of being more than simply constituting a larger external and historical 'context' within which these phenomena are subject to change.[10] Like a material myth that keeps telling itself, each talisman is comprised of an

---

[10] To be sure, the present account of Mongolian shamanism hardly qualifies as 'historical' according to the established genre-criteria of 'historical anthropology'. For this to be so, we would need to consider the 'larger' political-economic context of postsocialist Mongolia and its impact on the shamanic 'micro' scale (see Pedersen 2011; cf. Geschiere 1997; Comaroff & Comaroff 1998). Yet, as Lévi-Strauss argued in his famous critique of Sartre (1962), humans engage with or indeed *do* time in

ever more heterogeneous bundle of materials stitching disparate events together across time.

Note how, as the aforesaid analysis proceeded, the talismans' bundle-like design became still more inseparable from the notion that the defining characteristic of the shamanic sprits lies in their capacity to always mutate from one form into another. In fact, had these artefacts not been imbued with the specific material properties that they are, it would have been hard or perhaps even impossible for us as anthropologists (and possible for local people too) *to think* of spirits as inherently labile entities. Let us try to pinpoint precisely what it is that we learned about the shamanic spirits/cosmos by 'thinking through' their talismans that we didn't (or, more strongly, *couldn't*) surmise from the more standard cultural exegesis with which we began our account. Providing a conventional ethnographic interpretation of local peoples' ideas about and practices around shamanic spirits, we started out by establishing that *ongod* are multiple phenomena irreducible to a single form. We then explored how the material characteristics of *ongod* might help to explain better this irreducible multiplicity and thus solve the paradox that a single 'transcendental' guardian can assume the shape of many different 'immanent' helpers. And indeed, homing in on the materiality of talismans allowed us to specify the way in which shamanic spirits are multiple, namely by virtue of the fact that any given *ongon* is comprised by a manifold of events (misfortunes, curings and offerings), which take place at

a qualitatively different way than through 'history', namely myth. Indeed, as we saw, it is precisely due to their 'myth-like' form that *ongod* are so to speak distributed *across* (as opposed to *in*) time. Far from constituting an 'extremely synchronic' approach (Ellis 2015: 56), then, our ontological analysis involves a *heightened sensitivity* to the temporal dynamics of socio-cultural life (see also Nielsen 2011, 2014). This is not to deny the importance of so-called broader historical and political processes in the analysis of shamanism – indeed, Pedersen's monograph on this topic (2011) was an explicit attempt to render different scales of analysis, ranging from micro to macro, as contexts for one another. But it is to question the hegemonic role that so-called 'larger contexts' often play in anthropologists' 'contextualizing moves' (Dilley 1999), as if other dimensions – such as, say, shamanic spirits – could not just as well be used to frame a given ethnographic material and its analysis.

different times and places and therefore cannot, by definition, be conceived of in the singular. Not unlike the manner in which the power of powder was adduced directly from 'motile' material qualities in the aforementioned analysis if Ifá divination in Cuba, then, the mutability of shamanic sprits was elicited from the strikingly heterogeneous composition of the Mongolian talismans. In other words, had it not been for the material properties of these artefacts, we would not have 'sharpened' the generic concept of multiplicity into the more precise concept of mutability. By enabling a conceptual transformation from 'multiplicity' into 'mutability', our experiment with the theoretical possibilities latent in the materiality of shamanic artefacts added a new dimension to our analysis of spirits and the shamanic cosmos writ large, namely the dimension of time, and by implication that of change, including the ontological question of what change *is* in such an ethnographic situation.

This, then, is the second manner in which these artefacts can be described as talismans of thought. The point is not only that the ever more diverse bundle of things of which talismans are composed adds up to a material database of spirits' past manifestations, which plays the role of a powerful 'mnemonic technology' (Küchler 1988) or indeed 'cognitive scaffolding' (Pedersen 2007; cf. Mithen 1996; Clark 2008) for local people. The point is also that these talismans, in light of what they can and cannot make visible, offer a unique *theoretical* insight into the nature of spirits and shamanic cosmology more generally. And for whom does this apply – whose 'thoughts' are these artefacts 'talismans of'? Clearly, the answer will first of all have to include Pedersen himself, as well as other anthropologists, including ones from Mongolia (e.g. Buyandelger 2013; Bumochir 2014). Much like the distinct material properties of powder allowed Holbraad to solve his ethnographic riddle of how *aché* powder can also be power in Ifá divination, so the characteristically heterogeneous form of Mongolian shamanic artefacts prompted or even compelled Pedersen to interpret his ethnography about shamanic sprits in a certain way. These talismans provided, in other words, the affordances for his anthropological conceptualizations. It is hard to overestimate how much the above account of Mongolian shamanism hinged upon the material properties of the artefacts in question. Without detailed attention to the strikingly and very deliberate heterogeneous design of the talismans several vital steps in the aforesaid line argument simply could not have been made.

So this is the sense in which the shamanic costume can be described as 'a theory that stands for itself' (Pedersen 2011: 181). As a cultural artefact that appears to contain within itself everything that is needed to explain it (cf. Riles 1998), the gown worn by the performing shaman does not need to be interpreted in a conventional, hermeneutic sense of the term, neither by anthropologists not arguably by local Mongolians, for this object and its specific materiality and manner of display already constitutes an explanatory context in and of itself. These artefacts 'analyse themselves', as their material properties set in train conceptual trajectories in their own right, which offer anthropologists (as well as local people, albeit in a different sense[11]) with

[11] While, as we have explained, this is not our primary concern in the present argument and in this book as a whole, a similar (but not identical) point can be made with respect to local people's conceptualizations. Certainly, Pedersen's ethnography leaves the sense that, were it not for these artefacts and their capacity to make the spirits (in)visible, also his Mongolian interlocutors' knowledge of the shamanic cosmos would have been different and possibly impaired. Indeed, there are reasons to believe that talismans were crucial transmitters of shamanic knowledge during seventy years of Communist repression when Mongolian shamans were stigmatized, and often murdered (Pedersen 2011: 119–22; see also Buyandelger 2013). Perhaps more than any other feature of the shamanic universe, it was the continual presence and peoples' use of these artefacts that made - and still today makes - it possible to be a client, and indeed practitioner, of 'shamanism' without being in possession of an elaborate body of knowledge about the spirits and the shamanic cosmos as a whole. After all, a distinctive feature of these artefacts seems to be that they do not require an extensive repertoire of shared meanings (a shamanic 'cosmology' or 'culture') on the part of the people who engage with them to work and make sense. Instead of being carriers of deep symbolic meanings – which is how religious talismans have often been analysed by anthropologists – these peculiar bundles of heterogeneous substances comprise everything that one needs to know about the

not just knowledge about but also a peculiar kind of theorization of the shamanic cosmos. In comprising all the dimensions or 'scales' necessary to be contextualized and understood, these artefacts can be described as postplural abstractions in the Strathernian sense. Their analytical potency does not reside in a capacity to generalize over many particular phenomena or instances (like ordinary abstractions), as we saw in Chapter 3, but in an obverse capacity to render things or events *more concrete* than they were before. In that sense, the talisman may be described as a sort of posthuman concept – a concept, that is to say, that was never made by anyone in particular, and certainly not within the confines of any single mind, but which nevertheless is imbued with a potency and a beauty that far surpasses most concepts invented by humans alone. Thus understood, certain material artefacts may be said to 'emit concepts' as radiation beams from radioactive materials, and in that sense may be subject to self-scalings, self-transformations and indeed self-abstractions. For is this not what a shamanic sprit 'really is': a material instantiation of the propensity of an always-changing cosmos to analyse itself? And what enables the *ongod* to perform this auto-analysis is their capacity to encompass everything there is and could be, not by hovering above the world as common abstractions are meant to do, but by always already being other to themselves due to the fact that 'change' is what they *are*.

shamanic spirits and the world of shamanism more generally. In this sense, to borrow a formulation from Strathern (1988), these artefacts may be conceived of as 'indigenous forms of analysis' in their own right – not, to be sure, analysis as anthropologists know it, but *what an analysis might have looked like* had 'anthropology' happened to have been a central concern of the people studied (more than, say, getting cured). Indeed, when Northern Mongolian shamans are possessed by spirits, this is all that people see: what they see is what they know. People do not have – let alone need – access to an 'underlying ontological regime' (Ellis 2015: 56) to partake in the ritual. All that they need to know is that each *hadag* knotted to a *manjig* is another misfortune/curing made visible, and all remaining cosmological components and dimensions fall into place on their own (Pedersen 2011: 180–2).

As well as talismans *of* thought, then, *ongod* may be conceived of *as* thoughts in their own right. But we need to proceed with caution here. For it should be stressed that, in considering this possibility, we do not want to say that these things somehow 'can think' on their own, with all the hornets' nest of problems that automatically follows from making this or cognate metaphysical claims (Morton 2013). Our ambitions are more modest (and much less metaphysical), for they amount to a critical-experimental exploration of what happens to anthropological concepts of materiality (and anthropological concepts of concepts, *sensu* Corsín-Jimenez & Willerslev 2007) when certain artefacts are treated as *sui generis* theories. After all, an important outcome of experimentally treating these artefacts as if they were imbued with a capacity for self-analysis or self-abstraction, is that the concepts of 'analysis', 'abstraction' and 'change' *themselves* inevitably become reimagined – reinvented, in Wagner's sense – as an integral part of the process. More than anything else (and certainly more than closet-philosophical ambitions), it is this prospect – namely, that new insights into the conditions of possibility of anthropological knowledge can be generated by posing ontological questions – that motivates us to carry out the present sort of analysis in the first place.

Unlike more metaphysically inclined posthumanists, our suggestion that shamanic artefacts are 'forms of thought' is thus always made in relation to specific knowledge practices and matters of concern, be they, as in the foregoing discussion, anthropological knowledge, local forms of knowledge (as intended in Pedersen's original TTT chapter), or indeed both matters of concern at the same time (as Pedersen arguably sought to do in his 2011 monograph). And much as Pedersen's TTT chapter was concerned with Mongolian as much as with anthropological concepts, so also the present discussion of conceptual affordances has developed in relation to local and anthropological problems. Talismans *as* thought, yes, but only when inseparable from certain kinds of shamanic and/ or anthropological concern. This, after all, is the signature move of the ontological turn: it is not a metaphysical treatise on what a given thing or phenomenon 'really is' (whether twins are really birds, powder is really power, or things really are thoughts), but a heuristic and inherently experimental exploration of what different concepts would have to be in order to have an anthropological purchase on the ethnographic materials they seek to elucidate and the analytical problems that these materials pose.

Pragmatology, or Art Backwards

By way of conclusion, we may return to our opening question about the status of things within the economy of anthropological analysis. For, notwithstanding our abiding emphasis on things' inherent capacity to transform themselves into concepts, which is the major tenet of our two examples, one may still wonder how far, and in what sense exactly, this counts as raising their analytical status 'as such'. After all, even though we have tried to disentangle it heuristically here, the complex entanglement of the things 'themselves' with what people say and do around them, illustrated in our two examples, shows the deep imbrication of things and people in the activity of ethnography. Whatever the merits of the case we have sought to make for things making a difference to our analyses of them, it would still seem that their status as objects of inquiry remains unavoidably circumscribed by the human-oriented agendas to which our analyses are directed.

Our Afro-Cuban example may be used to illustrate the objection: Sure, powder may be operative in the analysis of *aché*, providing the material source for concepts such as perviousness, multiplicity, intensive motion and so on. Still, these conceptual ingredients are deeply embedded in an overall argument about Cuban diviners – that is, people – which itself relies on a host of other ethnographic data, concerning not just things like powder, but also divinities, diviners, their clients and so on. While our conceptualizations might be driven (partly) by things 'as such', their anthropological significance is, after all, to be gauged with reference to the ethnographic conundrums posed by the people in whose lives the things in question feature. So the question about 'things as such' seems to remain: could things really dictate the terms of our analytical engagement with them without their association to human (in this case ethnographically talkative) subjects?

But here we can bite the bullet. Examples such as our analyses of Afro-Cuban powder and Mongolian shamanic artefacts indeed *do not* demonstrate that things can dictate the terms for their own conceptualizations *entirely* of their own accord, and seem bound to continue to render things' conceptual transformations in relation to the analysis of the human projects into which the things in question enter. The analyses to which the things as such make a difference, after all, are anthropological, concerned with people's lives. What these examples *do* demonstrate, however, is that such anthropological analyses can involve an irreducibly thing-driven *component*, or *phase*, and it is just this component that we have sought to isolate and clarify in this chapter with reference to the 'thing = concept' formula. In order to distinguish it methodologically from the ethnographic component of anthropological analysis, with which it may in practice be entangled as we have seen in the example, we venture to call this thing-driven manner of conceptualization 'pragmatological' (cf. Witmore 2009). Emphasizing the origination of thinking *from* (as opposed to just through) things, the term designates the activity of extracting concepts from things (*pragmata*) as a distinctive analytic technique.

Indeed, while in examples such as the ones we have provided the analytical difference things can make 'pragmatologically' is gauged strictly with reference to the anthropological mileage they afford, the very notion that things can make such a difference of their own accord, as such, also raises the prospect of pragmatology as a *sui generis* mode of inquiry – one that may feature in a variety of disciplines, beyond socio-cultural anthropology. For example, recalling our earlier discussion of Science and Technology Studies (STS), one could explore whether the entanglement between humans and non-humans that lies at the heart of STS methodology might also contain within it a pragmatological component. Transposing the fundamentally heuristic premise of our argument, we could then venture to isolate the conceptual difference that 'materials' make to STS analyses without thereby reinstating the ontological divide between humans and non-humans that STS so productively denies. Similarly, perhaps, one could explore the mileage pragmatological analyses might have in archaeology, a field in which the need to derive analytical insights from things in themselves, with characteristically minimal human input, is perhaps most acute. In this way our argument about pragmatology could seek to contribute to recent attempts to adumbrate the implications of the ontological turn for archaeology, most notably by Benjamin Alberti (e.g. 2014a and see Chapter 1).

Taking the thought a step further: could one try to imagine an independent, thing-centric discipline called pragmatology, in which things' material properties would form the basis of conceptual experimentations that would be unmediated by any human projects whatsoever? What might such a discipline look like? Ian Bogost has recently raised a similar question, although his answer is slanted more towards conveying the *experience* of conceiving things beyond human mediation, to develop what he calls an 'alien phenomenology' (2012). In light of our concern with the conceptual effects of things, perhaps theoretical physics would come closer to what we have in mind, since so much of it seems to take the form of radical conceptual experimentations in the service of understanding the material forms of the universe. Still, this also has problems, partly due to physicists' still encompassing demand for naturalist 'explanation' – a demand that would therefore need to be disentangled from our pragmatological concern with conceptualization (see Introduction). At any rate, there is no a priori reason to limit the comparison to physicists' takes on matter, to the exclusion of those of, say, chemists, biologists, engineers or, indeed, artists, sculptors or musicians.

Perhaps the best comparison would be with some form of abstract art. Speaking very broadly, might we say that the labour of the abstract artist consists in producing an object that congeals in concrete form a set of conceptual possibilities? As they are exhibited in a gallery, are Malevich's *Black Square*, Kandinsky's *Picture with a Circle*, or the grid-like forms that Rosalind Krauss finds recurring in works of twentieth century modernism (1986) not in some pertinent sense 'concretions' of concepts? If this is a fair way of imagining (at least some) abstract art, then pragmatology could be conceived as performing the same procedure in reverse. Accordingly, the pragmatologist's work of 'postplural abstraction' consists in producing conceptualizations that express in abstract form a set of concrete realities. Pragmatology, then, could be considered as a form of conceptual expression. Art backwards, even.

All this, of course, is to go out on a limb. Still, however indulgent, we would suggest that these kinds of speculations serve to illustrate the spirit (if not the kind) of experimentation that we see as a prime characteristic of the ontological turn. Indeed, our attempt in this chapter to transpose the core approach of the ontological turn onto the study of things is itself an example of the kind of methodological experimentation this manner of anthropological thinking seeks to pursue. By dwelling at the limit of the ontological turn, the engagement with 'things as such' that we have explored has spun the ontological turn itself in a new direction, from ethnography to pragmatology, in the very act of trying to extend it. In the next chapter we continue in the same spirit of experimentation to explore what happens when one sets up for reflexive reconceptualization one of the ontological turn's own basic premises, namely the idea of the relation. Where might the ontological turn go if it were to turn on itself?

SIX

After the Relation

Throughout this book we have seen that the concept of 'the relation' has been central to the distinct form of anthropological thought that eventually developed into the ontological turn (Wagner 1977b; Strathern 1988, 1995a, 2014a; Viveiros de Castro 2004, 2014). In the work of Strathern, in particular, the relation is explicitly identified as the indispensable mode of anthropological inquiry, while for Viveiros de Castro this is rendered as a (in fact, *the*) form of ontological transformation. In this final chapter we raise the question of what would happen if we targeted the radically reflexive and abidingly experimental attitude that, as we have argued, is fundamental to the ontological turn, at the concept of the relation *itself*. Can we think of an ontological turn 'after the relation'? And what might such a 'post-relational' move look like?

In posing these questions, we wish to contribute to a debate that has been picking up steam within certain quarters of mostly European anthropology over recent years concerning what might be the limits of 'the relation' as an anthropological concept and analytics (Corsín-Jimenez 2007; Rio 2007; Stasch 2009; Humphrey 2009; Candea 2010a; Pedersen 2012b, 2013b; Gatt 2013; Candea, Cook, Trundle & Yarrow 2015).[1] One of

---

[1] Among the handful of anthropologists who have explicitly explored what might come after the relation as a concept, Alberto Corsín-Jimenez's approach resembles our own most closely. 'The trouble with the Melanesian model', he suggests, is that 'proportionality [is] a constant, and it is therefore assumed that people always

the most concerted attempts to address this question has been made by Michael Scott. Indeed, one of Scott's main concerns in formulating his 'poly-ontology' (see Chapter 1) has been to counteract what he sees as the undue dominance of the 'Melanesian Model of Sociality', which he ascribes to Roy Wagner and Marilyn Strathern. He writes:

The concept of intrinsic relations or multiplicity – the idea that entities are relational internally as well as externally – is an indispensable insight disparately argued and collectively established within anthropology, especially by Wagner, Strathern, Latour, and Viveiros de Castro. Yet this insight has prompted an unwarranted assumption responsible for the theoretical asymmetry between relations and entities that I have found problematic since I first began working with Arosi ... Morten Pedersen has already asked 'what might come after the relation' [2012a: 69]. By this I take him to be asking what will succeed the nondualist meta-cosmology in anthropological theory? My answer would be that we need a meta-cosmology with double vision ... able to treat relations and entities as fully coeval and equally available as viable premises for both anthropological practice and indigenous ontology. (2014: 50)

Our aim in this chapter is to heed this call to explore what might come after the relation in its capacity as one of the most abiding and 'primordial' concepts of the ontological turn. Unlike Michel Scott, however, we shall not do this by formulating a 'double' theoretical perspective that is complementary or more precisely 'strabismic' with

relate in the same fashion ... A full model of proportional sociality ... is one that takes into account the different ways in which people inflect and qualify their relationships [... and] the factors by which the stretching out of the social takes place' (2007: 193–4). No matter whether this depiction of the Melanesian model is fair or not (Willerslev & Pedersen 2010), Corsín-Jimenez's call 'for a sustained focus on social and cultural practices of apportionment' is astute. After all, 'unlike relations, which only tell you how to disaggregate, apportionments tell you what to disaggregate into ... the apportionment is the form the relation takes when it emerges as a consequence; or to say it somewhat differently, when it works as a proportion' (2007: 187). For a more recent, and quite different, attempt to delineate and explore the ethnographic and theoretical terrain of a post-relational anthropology, see Candea et al. 2015.

respect to the theoretical approach developed by the ontological turn.[2] On the contrary, we are going to show, for such a 'post-relational' analytics to remain faithful to the impetus towards radically reflexive conceptual experimentation we take to be the hallmark of the ontological turn, it should not involve an attempt to moderate or roll back the kind of relational thinking spearheaded by Roy Wagner, Marilyn Strathern, Eduardo Viveiros de Castro and indeed Bruno Latour. Accordingly, as we shall argue, to ask what comes after the relation means intensifying the relational analytics associated with the ontological turn by experientially exposing the concept of 'the relation', and the analytical methods founded upon it, to ethnographic phenomena that have not hitherto been the subject of this approach – phenomena, that is to say, which may be designated as 'apparently non-relational'. As such, the post-relational project can be described as an attempt to stretch or extend the concept of 'the relation' a notch further than first the structural functionalists and later Strathern did (Chapter 3), and irreversibly 'reinventing' or 'obviating' it (in Wagner's sense) in the process, – not unlike the way in which the postmodernists did not seek to reverse to a pre-modern or a non-modern position, but aspired to take the modern still further and thus in a sense become even more modern than modernism itself (Pedersen 2012b).

So, in transforming the concept of the relation by asking what might succeed it, we simply enact the prime task of the ontological turn, namely to experiment with ethnographically derived forms of anthropological reconceptualization – only in this instance, this experimental reconceptualization is performed on the ontological turn *itself*. We have picked two candidates for our exploration of what might happen when

[2] By 'strabismic', Scott refers to his recent proposal for a necessary methodological indecision between relationism and essentialism in the anthropological study of ontological questions (2014). Thus, for Scott, ontologically inclined anthropologists need to adopt a 'dual vision' due to the fact that relationalism and essentialism cannot be reconciled in a unified theory without always giving relations the upper hand.

the concept of the relation is extended towards non-relational ethnographic materials and contexts. The two are closely related in the sense that they are implicated in each other and serve as the ground for one another: namely, the notion of the transcendence of God in Christianity, and the experience of converted Christian subjects. It should be obvious why these are particularly apt candidates for conducting our post-relational experiment. If there were ever two objects of ethnographic inquiry that did not seem to be relationally constituted, then it is a transcendent God and the experience of conversion – both seemingly implying an interruption or even a *break* of relations by definition. Our task, then, is to treat the concept of the relation as a baseline ripe for further transformation when confronted with ethnographic material that seems to cut against it, viz. the apparently non-relational characteristics of Christian faith.

In what follows, we shall not attempt to perform this post-relational shift fully, but only to sketch out, gesture-like, what it might entail. After all, as we have argued throughout this book, the major premise of the ontological turn is ethnographic: its reflexive project of experimental conceptualization is founded in and precipitated by the contingencies of particular situations and the analytical challenges that their subtleties pose. Entering into the subtleties – not to mention the notorious complexity – of particular empirical cases of Christian worship is well beyond our present remit and indeed our expertise. Our intention, rather, is strictly programmatic and experimental: we wish tentatively to demonstrate what a post-relational analysis might look like in practice. What follows, then, is an illustration of an analytical strategy, which can only be provisional because we are presenting an argument that is inherently about its own movement, if you like – an argument about where the ontological turn might go, were one to reconceptualize reflexively its relational method, and as part of this post-relational move perhaps also transform and alter its own theoretical premises (which by the same token are also recognized as inherently temporary, always contingent on particular ethnographic exposure, and therefore constitutively provisional). To an even stronger degree than was the case in the previous chapter, then, we are asking the reader to bear with us, reading this chapter, not as a fully fledged analysis of Christianity, but rather as an attempt to use the challenges that Christianity poses as an impulse towards exploring one of the ontological turn's prime conceptual frontiers and how these might be pushed further.

Christian Conversion and the New Melanesian Ethnography

Ever since the publication of *The Gender of the Gift* (1988) Melanesianist anthropologists, along with scholars working in other parts of the world, have sought to engage critically with the ethnographic and theoretical assumptions of the so-called New Melanesian Ethnography (NME) that Strathern's book instigated. Indeed, the limits of NME have been at the heart of one of the most heated debates within the subfield known as the 'anthropology of Christianity' (cf. Cannell 2006; Bialecki, Haynes & Robbins 2008). While reviewing this extensive literature is beyond our remit, here we focus on a part of the debate that is especially germane to our present purposes. We are referring to ongoing discussions among especially Melanesianists anthropologists, such as Joel Robbins (2004a), Mark Mosko (2010) and Eric Hirsch (2008), but also Amazonianist (Vilaça 2011) ones, about whether the 'relational' analytics associated with Strathern and other NME proponents provide an adequate theoretical framework for understanding what happens in processes of Christian conversion, including the question of whether or not conversion in Melanesia and other parts of the world involves a discontinuous rupture from non-Christian to Christian moral economies and forms of selfhood. At the heart of this debate is the question of 'the relation' – of its precise meaning as an anthropological concept and its ethnographic scope and traction when it comes to conveying the often dramatic societal and existential transformations that tend to accompany processes of conversion.

The foremost representative of this debate is Joel Robbins. In a number of influential publications based in his fieldwork among the Urapmin people of Papua New Guinea he has put the case for an anthropology of Christianity that takes more seriously how converts themselves experience, or talk about the experience, of conversion, namely as a radical, discontinuous, and irreversible rupture from one religious, cultural and moral order to another. Referring explicitly to Louis Dumont but harking also more implicitly back to other influential theories of modern individualism and Christian (Protestant) selfhood including those of Weber (1958), Robbins writes:

Melanesian cultures value the creation of relationships over that of other cultural forms (e.g., individuals, wholes) and ... reckon the value of relationships rather than individuals who make them up or the larger structures of which they may empirically be a part ... With the coming of Christianity, the traditional relationalism of Urapmin culture has been severely challenged. This is because Christianity is unrelentingly individualist ... This individual before God is the paradigm of 'the independent, autonomous, and thus essentially non-social moral being, who carries our paramount values' ... [W]ith Christianity's arrival, relationalism has lost its right to occupy without question the paramount slot in Urapmin culture. (2004a: 292–3, citing Dumont 1986: 35)

For Robbins, then, the NME model reaches an unsurpassable analytical limit when confronted with ethnographic realities, such as Christian conversion (and other vehicles of modernity), that, he asserts, cannot be deemed 'traditionally relationalist'. Of course, as Robbins points out in another influential publication (2007), 'anthropologists assume that people's beliefs are difficult to change and therefore endure through time' (2007: 139). Yet, this preference for 'continuity thinking' in anthropological analysis is empirically challenged by the fact that, in Melanesia and everywhere else in the world, 'Christians ... tend to imagine their religion as historically constituted by Jesus' rupturing of earthy time by his birth ... [and] expect such change to occur in their own experience at conversion' (2007: 13). And nowhere, maintains Robbins, is this transformation from 'relationalism' to 'individualism' expressed more clearly than in confession. This is so, 'because a confession affects only the person who makes it and the person's relationship to God, one cannot confess for another. Thus disengaged from the relationships that make up the social world, the confessing subject in Urapmin is a perfect example of the "non-social" yet still moral individual of Christian individualism' (2004a: 297-8). In short, for Robbins and his followers, in order for anthropology to take seriously the individualism entailed in Protestant conversion (and Christianity more generally), they need to move away from the 'dividualism' associated with Strathern and the NME model.

But perhaps the NME model has been too quickly dismissed by Robbins and other scholars (Carrier 1992a; Martin 2013) who have focused on its analytic and ethnographic limitations. This cer-tainly is Mark Mosko's message in a much-discussed paper (2010), which explicitly was motivated by what he laments as the ossifica-tion of this question into two 'deeply bifurcated' ethnographic and theoretical camps:

Just as [NME-associated anthropologists] deploying the partible person model have been inattentive to Christian missionization and conversion, studies of the latter have presupposed Christian personhood to be strictly individualist ... I ... seek to overcome this specific impasse by examining several well-documented cases of religious transformation [where] the sort of 'individualism' routinely attributed to Christianity actually consist in just one aspect of a wider encompassing unbounded form of personhood – related to, but distinct from, the bounded possessive individual – which is instead closely analogous to Melanesian dividuality ... Suitably reconfig-ured around the dynamic potentialities of personal partibility, therefore, the [NME model] offers to social anthropology novel theoretical insights into Christianity and processes of religious change as well as continuity. (2010: 217–18)

Christian Conversion and the New Melanesian Ethnography

This is not the place to engage in detail with Mosko's bold but controversial (Robbins 2010; Barker 2010; Scott 2015a: 648–9; see also Vilaça 2011; Mosko 2015) attempt to extend the NME model to Melanesian converts, who, on his interpretation, are 'Christian dividual persons' in the Strathernian sense, emerging from 'the conversion of one dividualist form of personhood, agency, and sociality into another' (Mosko 2010: 232). Suffice it to say that Mosko is suggesting that Christian Melanesians are just as 'relationally' constituted as non-Christian Melanesians, for, rather than constituting an 'independent, autonomous, and thus essentially non-social moral being', as Dumont and Robbins would have it (cf. Robbins 2004: 292), 'individuality' on this account is simply one enactment of personal dividuality and relational partibility among many. As Mosko puts it, the 'inner, indestructible and sacred but indivisible soul' of the converted (in)dividual is 'only one component of the plurally constituted total Christian penitent', who 'comprises at minimum inner immortal soul, will, and faith as well as a mortal body, outer devotions to God, good works directed to fellow communicants, and so on' (2010: 220).

At first glance, Mosko's goals look identical to ours, namely to extend the relational analytics associated with New Melanesian Ethnography into uncharted ethnographic and theoretical territory. And also like us, Mosko takes Christian conversion to be the perfect test case for carrying out this analytical experiment due to the marked individualization that is widely reported to be one of its key sociological outcomes. Still, the problem with this kind of analysis is that it seems not to correspond with how many Melanesian Christians *themselves* describe the process of conversion, namely as a radical break with the cultural norms and moral values of the past; a rupture that involves highlighting the individual agency of each believer and her relationship to God at the expense of heterogeneous relationships with a plethora of human and non-human agents, which Christian converts are so deliberate about leaving behind

(Barker 2003).[3] Indeed, it is hard to dismiss Robbins's take on conversion as the coming into being of less socially embedded persons; after all, this is just how many converts across the world depict the process of becoming a Christian, namely as involving a reduction of human and non-human relations (see Keane 2007: 52, on the 'anti-social demand made by many Protestant missions' in Indonesia; see also Meyer 1999).

To be sure, not all Christians think of conversion as a radical break with the past. As Michael Scott puts it in a recent critique of the 'rupture model' associated with Joel Robbins and his associates, some scholars 'tend to construct only one model of personhood ... and they posit this model as definitive, either of Christianity as a whole or, more narrowly, of the particular kinds of Christianity they study' (2005: 636). Yet, he goes on to say, 'I would not deny that there are Christians who think and act according to various versions of a quasi-atomistic, vertically oriented model of Christian personhood. I would argue, however, that such a model is not alone in the history of Christianity; others have always coexisted, either in combination or conflict with it' (2015a: 638; see also 2007: 302–5). Thus, the argument about conversion that we are trying to make here may not be equally relevant to all Christian contexts. Having said that, the ethnographic record is quite clear that questions of irreversible social and existential rupture lie at the heart of Christian conversion for many different peoples across the world. As Robbins, Schieffelin and Viveiros conclude in a comparative study of Christian conversion in Melanesian and Amazonian contexts,

at least in our three cases there does appear to be a 'logic' of Evangelical selfhood that is making itself felt in otherwise diverse processes of conversion.

---

[3] In addition to this and other ethnographic reservations about Mosko's argument (Robbins 2010; Barker 2010), his 'NME revisionism' seems to suffer from potential theoretical problems too. For in suggesting that the 'dividual' is an aggregate of detachable things (like sin, faith etc.), which is anchored on an impermeable and indestructible atomistic essence called the soul, the dividual ceases to be a relational concept in the (postplural) Strathernian sense and appears instead to become reduced to a small-scale pluralism. (We thank Michael Scott for this point; see also Scott 2015a.).

This is a logic that ties a growing emphasis on the inner self to a devaluation of the bodily contribution to selfhood. This focus on the inner self is further linked to a decreased (though never wholly absent) moral interest in the state of social relations in favor of one placed on the inner self, particularly as it is known by and related to God (2014: 586).

It would appear, then, that the discussion about Christian conversion in Melanesia has reached a stalemate of sorts. How to account both for the fact that (if we are to follow Robbins and other ethnographers) Melanesian converts stress their irreversible transformation from socially embedded persons into autonomous individuals when becoming Christians, *and* the fact that (if we are to believe Mosko and other scholars, including Robbins himself, 2004a) Melanesian social life is otherwise best understood in relational terms (see also Hirsch 2014)? If we go for the first option, we gain the advantage of taking our interlocutors seriously as Christian believers but seem to lose the scope for ethnographic particularity and cultural variation in the process; yet if we are to stay within the NME approach in our study of conversion, we seem to seriously close down the possibility for any sort of radical change, be it social, religious, cosmological or existential, taking place among the people studied.

A possible solution might be to adopt the 'poly ontological' perspective proposed by Michael Scott (2007) or the related approach suggested by fellow Melanesianist Knut Rio (2007).[4] This would entail a classification of the social forms under investigation into opposing but coexisting cultural logics, cosmologies or ontologies in Scott's own sense (see Chapter 1). In the case of Christian conversion, this would

---

[4] According to Rio, 'the relation as a solitary analytical apparatus for discussing sociality quickly limits itself to a narrow vision of the social' (2007: 27). Thus, we need another perspective that is complementary to 'ideas concerning the relation, which come specifically out of New Guinea social ontologies ... [a perspective which] ... enables us also to discuss what exists socially on the outside of relations' (2007: 27–8).

be a matter of distinguishing what Robbins calls the 'relationalist' logic of non-Christian Melanesian persons from the 'individuality' logic of Christian Melanesian selves, and showing how the two coexist in more or less mutually enforcing or destructive combinations. Clearly, this is a viable and well-tried option that has resulted in compelling studies of the negotiations, mediations and contradictions that navigating between conflicting moral orders entail, including Robbins's *Becoming Sinners* (2004a). But there is also another, and in a sense opposite, path out of the stalemate, which we shall pursue here. What if instead of looking for a schism between relationalism and individualism (Robbins) – or for that matter searching for different forms of coexistence and complementarity between 'polyontoly and mono-ontology' (Scott) – we try to change the two poles themselves by re-inventing the core concepts on which each of them respectively relies, namely that of the relation and that of the individual? What if, rather than distinguishing relationism from individualism, we were to expose the analytical schemes founded on the former to the ethnographic contingencies identified as the latter, in order to see how relationism might be *modified* by its exposure to the ethnography of the individualism of Christian conversion?

In what follows, we sketch the contours of what such an alternative theoretical account of Christian conversion in Melanesian and other contexts could be. We do so by formulating an answer to what we consider to be a post-relational question, namely what happens to all the social relations that seem to 'disappear' in conversion. By exploring what happens when these 'missing' relations are heuristically assumed to re-appear on the inside of the 'swollen self' of the born-again individual, as we shall depict it, we hope to show not only how the 'individual' can be turned into a relational concept (after all, Strathern already did so with her appeal to the notion of a 'dividual'). Above all, we hope to show, this raises the post-relational question of how the concept of the relation *itself* might be inflected by the concept of the individual in such a way, crucially, that it ceases to be about 'social relations'. To explore these questions, we now present a case study of conversion carried out by Pedersen in Mongolia's capital Ulaanbaatar.

The Great Indoors: A Case of Christian Conversion

While most Mongolians classify themselves as Buddhist and/or shamanists, the number of Christian converts has been increasing rapidly since the collapse of state socialism in Mongolia, particularly in the cities, and especially among younger, educated women (Pedersen 2012b, Højer & Pedersen forthcoming). Consider the case of Undraa, the oldest unmarried daughter in a large household of rural migrants settled in the capital of Ulaanbaatar's peri-urban slums. At the time when Pedersen got to know her around 2000, Undraa was struggling with a number of problems with drunken and jobless men, who constantly made new claims (money, alcohol, sex) on her as well as other women around her without giving anything but trouble and occasionally violence in return. More generally, Undraa had always found it hard to deal with what she considered the excessive demands and troubles inflicted on her by people in her vicinity, ranging from pushy customers during the period where she managed a kiosk for her family, to when newfound acquaintances (*tanil*) impatiently sought to turn transient relationships into enduring friendships. Partly in response to this dissatisfaction with her surroundings, and partly due to the fact that she had always been looking for a meaning in life that had hitherto escaped her, as she explained to Pedersen in retrospect, Undraa began frequenting one of the many new, mostly Evangelical Protestant, churches that had opened in Ulaanbaatar since the 1990s.

Eventually, Undraa converted, significantly, at about the same time as she opened a small business. Both changes were gradual and incremental as opposed to sudden and abrupt: it took time to obtain the required permissions, and necessary resources, to set up the company; much in the same way as Undraa's faith took years to mature in her, reflecting as it did a concerted effort on her side to study the Bible and try to understand the word of God, the sacrifice of Christ, and the omnipresence of the Holy Spirit. Nevertheless, taken together, they were also events that gave rise to a radical and irreversible shift in Undraa's make-up as a social, moral and spiritual person. For her double shift to a new religion and a new occupational status was accompanied by a hardened rhetoric in response to some of the claims made on her by different people around her, ranging from potential suitors to recent acquaintances who wanted to become part of her life. Eventually, around 2012 when Pedersen last saw her, Undraa's life had become stabilized, or one might even say frozen, in the same pattern: her business was prospering (without making her rich) and her relations with various (male) relatives and acquaintances was still strained without having been completely severed.

Meanwhile, Undraa's attitude to the plethora of spiritual entities comprising Mongolia's shamanic cosmos (Pedersen 2011; Buyandelger 2013) also underwent a dramatic change. Like so many other young Mongolians during the chaotic years of postsocialist transition (Pedersen & Højer 2008; Højer & Pedersen in press), Undraa had spent much of her adult life years seeking in vein to establish a safe distance between herself and what she perceived as the omnipresent spirits encroaching on her from all sides, not the least during the many long nights when worries about the well-being of her family, or the accusations made by friends, prevented her from sleeping. Like several other new converts that Undraa eventually befriended after joining the Church, she had therefore been searching desperately for ways to suppress a perceived over-abundance of 'black' (har) forces, which were so diffuse and manifold that she would sometimes find her mind and body overtaken by 'darkness', as she put it.

Indeed, it soon became clear that Undraa's conversion to Christianity to a large extent was the result of her own experiences with 'evil spirits' and other harmful 'influences' (nölöödöl). Nevertheless, and much as in the Urapmin case described by Robbins (2004a), this did not lead her or

other Mongolian converts to whom Pedersen spoke about these matters to begin doubting the existence of these non-Christian spirits
(apparently, this is a common phenomenon among charismatic and
Pentecostal converts, cf. Robbins 2004b: 128–9; Chua 2009). Only
now the difference from before was that, after having found God and
put all her trust in Him, Undraa did not fear the shamanic spirits in
the same manner as she had before. By constantly working hard to
remain detached from dangerous spirits and obnoxious men Undraa
was able to substitute her vulnerability to multiple harmful shamanic
or demonic influences with an unquestioned love for a single God.
Having fundamentally re-organized – or, to borrow a term from
Corsín-Jimenez (2007), 'reapportioned' – her relations with different
human and non-human others, becoming a Christian enabled Undraa
to utter, with equal moral conviction and social efficacy, the two seemingly unrelated speech-acts 'I do not believe in the spirits' and 'no, I
don't want to give you money for vodka'.[5]

For Undraa and other converts in Ulaanbaatar, then, the dual transformation to a new (Christian) faith and a new (middle) class status
was, above all, a question of how to cut connections. Certainly there is a
revealing correspondence between Undraa's and other young, upwardly
mobile women's wish to absent themselves from the excessive claims
made by friends and family on the one hand, and their attempt to free
themselves from multiple spirits' influence by renouncing and avoiding
them. Thus, Christianity seems to provide an escape from two negative

[5] As Undraa on several occasions told Pedersen, when Mongolians told him (as
many people did) that they 'didn't believe' (idgehgüi) in spirits, this did not mean
that they doubted their existence. On the contrary, she stressed, people expressed
disbelief in spirits precisely because these entities were so much on their minds
that they wanted to reduce the spirits' capacity for inflicting harm via, say, acts of
witchcraft or sorcery (see Højer 2009). 'So you see,' Undraa instructed Pedersen
in no uncertain terms 'when people here tell you that they don't believe in the
spirits, they do it because they don't want a part of the shamanic thing with curses,
counter-curses, and so forth. For if you don't believe in the shamanic spirits, then
you will also be left alone by them!'

influences at the same time. Indeed, this is what conversion essentially is all about, for Undraa and many other Christian converts in Mongolia: 'a ritual of rupture that drives a wedge between converts and the contemporary social world' (Robbins 2004b: 128). The convert severs a heterogeneous manifold of unwanted social and spiritual relations to fully and finally concentrate his or her trust and obligations around a single relationship, namely, the relationship between an individual human believer and an almighty God.

But this begs an interesting ethnographic and theoretical question that has not, it seems to us, been fully addressed by the anthropology of Christianity. We are referring to the unresolved question of what happens to all the social and spiritual connections severed in processes of conversion – where, as it were, do all the cut-off relationships go? While processes of Christian conversion have often been understood by anthropologists to involve a reduction of relationality (sociality),[6] it might, we suggest, be more analytically interesting and potentially more faithful to the ethnography to think of conversion as a *relational transformation* in Strathern's sense. Crucially, by referring to this move as Strathernian, we want to highlight the fact that 'relations', or for that matter their lack, are not something that there is more or less of 'out there' in particular ethnographic realties that are independent from anthropological concerns, but are rather conceptual operators that are internal to particular modalities of anthropological thinking, as we discussed in detail in Chapters 3 and 4. Christian conversion, we are going to argue in what

---

[6] Note how this line of argument relies on an identification between the concepts of relationality and sociality: to relate is to be social, and vice versa. While this 'fetish of connectivity' (Pedersen 2014) – where connections between people (and between people and things) are treated as ideal relations – is tacit in much sociological and anthropological theorizing, it is not always present in the work of leading anthropologists of Christianity. For example, in his more recent writings, Joel Robbins makes it clear that there are not necessarily less relations 'out there' after conversion; the point is rather that these relations are valued differently (see Robbins 2015).

remains of the present section, may thus be theorized as an enfolding or involution of human and non-human relations that keeps relational complexity intact, but displaces this complexity from an exterior realm we tend to think of as 'social' to an interior domain we here shall call 'existential'.

To substantiate this point, let us return to the story of Undraa. Evidently, for her being a Christian is hard work. The effort she puts into resisting urges – like when she begins the day by renouncing spirits that appeared in her dreams, – confirms that, for her, the primordial rupture of becoming Christian must be repeated over and over again. Indeed, she never seems fully satisfied with her efforts, either in the context of her work (although she works long hours), or when it comes to her practising her faith (despite being a regular church-attendant, she always questions her commitment to God). In fact, the more she tries to believe in God and express her loyalty to him through outward ceremonies and celebrations and inward prayers, the greater her doubt also seems to become; not doubt about God's existence, but doubt about whether she, as merely human, does enough to deserve God and his unlimited grace and love.

Clearly, for Undraa (as for many other Protestants, Ruel 1982), the constant affirmation of belief (and doubt) is at the heart of what it means to be a Christian. This observation is corroborated by the time it took before she found herself able to declare to her family, friends and congregation, or even herself, that she had converted. Following a protracted period of time where she had attended Church, read the Bible and discussed with the congregation and its pastor, Undraa found herself ready to 'say out loud that I believe in God' and 'tell my parents that I love them and apologize for the past'. It was, she stressed, an important day, more important than any other in her life, and she had rented a conference room at a fancy hotel for the occasion. Everyone dear to Undraa was present, but most important of all visitors were her parents, who, as guests of honour, received a specially heartfelt thank-you speech delivered by Undraa as an integral part of the moment she and everyone else had been waiting for, namely the emotional address to the guests in which Undraa made it known to all present, humans and non-humans, that she was now a Christian believer and a child of a God.

It is no coincidence that Undraa hesitated so long before finding herself ready and worthy to announce her faith in God, her parents and everyone else. Indeed, it is hard to find a person more systematically self-scrutinizing and self-critical than Undraa. Over the fifteen or so years that he has known her, and especially following her conversion to Christianity, Pedersen was often left with the impression that much of what she considered to be the most significant events in her life took place inside her, in what might – with a nod to Meillassoux (2008; see Chapter 1)- be called 'the great indoors' of sustained solitary introspection (cf. Rapport 2007; Pandian 2010). That Undraa was not alone in thinking like this and in wanting to communicate it to others became clear in an interview with another female convert. Asked by Pedersen where the Holy Spirit (*ariun süns*) might be located – inside believers or outside them – the woman responded:

Inside, inside, inside (*dotor, dotor, dotor*)! Because I feel that it's inside when I talk to God or pray for something. Exactly inside, faith starts exactly on the inside. When faith grows, the inside gets bigger. It has something to do with the Holy Spirit. We pray and God fills us up with Holy Spirit from the top of our head to end of our toes. We want to live like God, but we can't actually be God. But because we want to live how it's written in Bible and live like God to fill ourselves with Holy Spirit, we want to grow bigger. If there is little Holy Spirit here in my heart, it will cover my whole body. Covering the whole body means we will be like God, give everything to God and live God's life.

As Webb Keane has shown in his study of Calvinist conversion in Indonesia, 'the work of purification that links Protestantism to the idea of modernity ... includes a special privilege accorded to individual's agency, inwardness, and freedom and a vastly expanded vision of the possibilities for individual self-creation ... that link agency to introspection' (2007: 54–5; see also Asad 1993). And crucially, this elevation of introspection as a privileged form of moral and epistemological agency happens at the expense of prior social relationships to different human and non-human agents; not necessarily in the sense that relationships to other people (and, as we as saw, to other spiritual beings) ceases to exist or matter, but because such exterior relations from now on seem to become inflected and encompassed by an interior relation to God. There is a sense, then, in which conversion involves a Strathernian process of figure-ground reversal where 'outer agency' swaps place and status with 'inner agency'. Or, in Scott's words, 'in conversion, the Holy Spirit calls people out of participation in sin, thus individuating them to some degree. But it does so in order to bring them into new relations that are not merely social but ontologically transformative, new relations of participation in the very being (body) of Christ' (2015a: 646). To be sure, the point is not that there are necessarily fewer social relations than before, or that social relations cease to matter. But the point is that these relations are modelled on and measured against the internal self-relations. Indeed, Undraa clearly seemed to conceive of 'social' relations as secondary in the sense of being inferior to the ideal self-self as well as self-God relations from which they were perceived to derive. As she explained to Pedersen:

As the inside grows bigger, there will be an effect on the outside. This effect will be shown in our actions. If I have little faith and little Holy Spirit, then I won't help people that much and I will find it hard to honest. But when Holy Spirit fills up in me and gets bigger inside me, I will have more faith and as soon as I have very strong faith, I won't do dishonest things. Something like this. People [missionaries] come from America and people [missionaries] who are from Mongolia go to the countryside and work among shamans and Kazaks [Mongolia's Muslim minority]. When [these people convert] their inside part gets bigger, they forget about themselves, have more heart for people. Their actions change, they may leave their work, change their lives, change everything. It is very big change in life, right?

It is this 'swollen inside' that emerges as the convert with Undraa's expression is 'filled up with Holy Spirit' that we propose to take more seriously ethnographically by suggesting that Christian conversion is a process of relational transformation, and not, as Robbins and his followers come close to, describing it as a process of relational reduction. Perhaps, when someone becomes a Christian, her 'self' does not just become more singular, bounded and discrete, as received wisdom has it. Rather, what happens might also be theorized as relational 'involution', where a multitude of exterior agents within the surroundings (spirits, drunken men etc.) is substituted by a multitude of interior affects (belief, doubt, love) appearing on the inside of a new, correspondingly inflated self. For is it not in this 'great indoors' of Christian introspection that all relations cut-off in conversion may be said to re-emerge, only now in what is a non-social, introvert form? And how otherwise to account for Undraa's conversion than by thinking of her as a 'self-relational' subject, who has converted social relations with external others into internal non-social relations between herself and a Holy Spirit within an existential interior that is imagined to have somehow swelled up in the process of conversion?[7]

Becoming Christian, on this interpretation, does not only amount to a process of social detachment and a celebration of individual autonomy by declaring one's loyalty to a single God, even if it does all those things too. Conversion should also be understood more literally as the transmutation of one form into another form, like when a convector turns steam into hot water. Thus understood, the converted self might be imagined as

[7] No presumption is made here that a concept of an 'inside' – and by implication a concept of a corresponding 'outside' – necessarily exist prior to the discourses and the process of conversion. As Robbins, Schieffelin and Vilaça suggest in their comparative study of conversion in Melanesian and Amazonian contexts (2014), it is possible to argue that in many cases Christianity creates the very inside it then populates.

a non-Christian relational cosmos that has, so to speak, imploded onto itself. Note how this 'interior swelling' (Nielsen 2012) is not the product of a relational transformation in the 'standard' NME sense espoused by Mosko. Rather, to understand what happens in Christian conversion is here recognized to require a full-scale conceptual re-invention, namely the merging of the concept of 'the relation' and of the concept of 'the individual' into a genuinely new concept: the 'self-relational individual'.[8] And, crucially, because the concept of the self-relation is qualitatively different from the 'social relations' that NME anthropologists (including Strathern) have tended to deal with, it can be described as the result of an inherently post-relational theoretical move. Becoming a Christian, then, does not merely entail a move of relations inside, but a change in the very nature of what a relation could be. Which is another way of saying that on this account, converting to Christianity is not so much a matter of doing away with relations, but a way of doing something different with, and *to*, relations: it is a matter not just of being relational in a new way, but of turning relationality itself into something that it was not before.

Before closing this first part of the chapter, let us try to specify the two central steps that were involved in making the preceding analysis. This will also clarify why it makes sense to call it post-relational and not 'just' relational: why setting up the problem in a particular way allowed us to formulate a more-than-relational solution to it, namely the post-relational

---

[8] Clearly, Foucault's account of the modern subject as a site of increasing self-reflexivity, self-monitoring and self-cultivation seems to resonate with the general line of argument suggested here. Indeed, Foucault sometimes referred to the introspective nature of this ethical work as 'self-relational' (Binkley & Capetillo 2009: 36–7; Taylor 2014: 128–9; cf. Foucault 1985), just as he explicitly traced its genealogy back to Christianity and its different transformations in the history of European thought (see also Asad 1993; Laidlaw 2004: 92–119; Scott 2015a). But, to our knowledge at least, he did not in any strict sense tie this interior relational 'swelling' to a concurrent reduction of exterior relations to human or non-human others by theorizing these new self-relations as a folded or 'obviated' external social relations.

concept of the 'self-relational' subject. The first step involved making the classic relational move, which we examined in detail in Chapter 3, namely to adopt Strathern's heuristic premise that 'at every level complexity replicates itself in scale of detail. The "same" order of information is repeated, eliciting equivalently complex conceptualization' (2004: xvi). In the present case, it was this 'postplural' move that allowed us to pose the question as to what happens to all the social relations (the complexity) that seem to be cut off in the processes of conversion. After all, as the standard NME argument would seem to have it, all this complexity (all these relations) must have shifted itself in another direction or onto another scale, only to reappear somewhere else in a new shape or form; it can't simply evaporate into the air, for complexity is here assumed to always remain stable no matter what the scale or indeed the context of the analysis happens to be (cf. 2004).

Now, for 'diehard' NME adherents such as Mosko, as we saw, the answer to this question would be to look for the missing relational complexity in what they stipulate as the fractally constituted personhood of the Melanesian Christian person, who is thus subsumed by the generic notion of a Melanesian dividual. However, due to the ethnographic and theoretical reasons outlined earlier, this was not the direction that we wanted to take in our analysis. Instead, our task was to take seriously the possibility that Christian converts really do undergo an abrupt transformation from more socially embedded to a more autonomous individual forms of self (*sensu* Robbins), but at the same time also explore what happens to our understanding of this if we try to stay within the analytical possibilities and constraints inherent to Strathern's relational analytical method. In doing so, we had to perform an experimental intervention on the existing scholarship on Christian conversion – we were essentially inviting the reader to join us in an anthropological experiment, namely the one made possible by framing our investigation within the specific heuristic parameters defined by what we have here called an post-relational analytics (see also Introduction). For, whereas the standard NME strategy has been to look for and identify new kinds of 'dividuals' in ethnographic phenomena and materials not previously encompassed by this theoretical model, such as the Christian self, what we tried to do earlier was subtly but crucially different. Instead of just seeking to widen the scope of the relational concept of 'the dividual' to encompass the non-relational concept of the individual, our account of the self-relational Christian individual involves a modification of the concept of 'the relation' itself.

This, then, was the second and distinctly post-relational step required to 'invent' the concept of the self-relational individual: deliberately to distort the Strathernian relational heuristics so that it ceases to always look for relational configurations involving social exchanges. Instead, the post-relational move hinges on granting oneself the distinctly non-Strathernian (because psychologizing) theoretical liberty to stipulate that the 'existential' introspections *within* a single person are just as complex ('relational') as the 'social' exchanges *between* two or more persons. The result is that the relational now ceases to be 'owned' by the social: the converted self emerges as individual *and* relational, at one and the same time. Christian conversion keeps relational complexity intact, but it changes relationality itself by causing interior self-self connections to proliferate at the expense of exterior social connections with human and non-human others.[9]

[9] It may be objected that, already from a conventional NME perspective, persons are understood to be both intrinsically and extensively relationally constituted. Thus, for Strathern, Melanesian 'dividuals' are composed of relations and nothing but relations, which are then either 'elicited' or 'eclipsed' in different 'aesthetic forms' in the course of social life (1988; see Chapter 3). Yet if this is the case, then how does the concept of the self-relational individual differ from more standard models of internal relations? For the reasons mentioned in the chapter introduction – that the point of the ontological turn is to conduct its analyses by means of concepts derived from ethnographic specificities – a full answer to this question can necessarily only be provided in a future, more comprehensive ethnographic exegesis of Christian

After the Relation

Several unresolved issues remain. For one thing, how might our account of the Christian subject as a 'self-relational' figure reverberate with more established psychoanalytical and poststructuralist 'relational' theories of the subject, such as for instance Lacan and Foucault? (See note 8.) Furthermore, one might ask exactly how the present approach differs from, say, the recent attempt to reinstate Gabriel Tarde's theory of monads to what Latour and others (Candea 2010b; Latour et al. 2012) take to be its rightful place at the heart of social theory and philosophy – after all, the self-relational subject would, on the face of it, seem to be a rather neat candidate for a theological monad in that Tardean sense. Nevertheless, as we hope to have demonstrated in this first part of the chapter, the concept of the relation has not run out of steam. For the moment that it is remembered that 'the relation' was never meant to be a feature of ethnographic localities and societies in Mongolia, Melanesia or elsewhere, but above all – indeed strictly – an internal property of the way in which a particular mode of anthropological thinking conducts itself (see Chapters 3 and 4), it also becomes clear why it makes sense experimentally to extend – and thus reinvent – it in the way we have illustrated here. In the process, we hope to have shown how such a move forward for the ontological turn can also gain traction on wider anthropological debates, such as, in this case, the one on the anthropology of Christianity.

Taking Transcendence Seriously

Having offered a tentative account of what a post-relational concept of the Christian convert might look like, let us now pursue a second and related

conversion in Mongolia. But we hope to have presented sufficient ethnographic specificity and theoretical reflection to point to the subtle but crucial difference between theorizing single persons a being composed by *the same kinds of relations* that are posited to exist between different persons (the NME perspective), and in theorizing the relations posited to exist within certain persons (Christian individuals) as qualitatively different from all other relations, including 'social' ones (the postrelational alternative that we have outlined in this first part of the chapter).

post-relational experiment with Christianity, in this case by seeing what happens if one tries to align the concept of the relation with the concept of a transcendent God. In this connection, rather than engaging with an ethnographic description of Christian practice, we shall conduct our experiment by way of a critical re-reading of Bruno Latour's attempt to render Christian cosmology 'flat' in a highly suggestive chapter advancing a thoroughly relational account of the constitution of Christian faith, as exemplified also in Catholic devotional art (2010a). Indeed, as part of the more encompassing concern with ontological symmetry that we have discussed in earlier chapters, Latour has advanced a series of sophisticated reappraisals of Catholic faith by contrasting it to his famous analysis of the practice of science (Latour & Woolgar 1986; Latour 1988). Elegantly, in the seminal study on which we shall be focusing here (2010a), he contrasts science and religion by placing them on opposite poles of an axis that lies at the heart of his conception of mediation, examined in Chapters 1 and 5 in connection to his relational model of 'actor networks', namely the axis that runs from proximity to distance. We begin by providing the rudiments of Latour's argument on Christianity.

Philosophers of science, suggests Latour, tend to assume that science is basically a sophisticated version of common sense, concerned with producing discourse that matches reality directly and transparently: there's the cat on the mat (common sense), and there's the neuron firing (science). Latour has for long insisted that this way of imagining how science works is prejudice – part of the 'purification' exercise of paring apart representation on the one hand and reality on the other that the 'modern constitution' prescribes. In action, science is a much more delicate business, whose truth and objectivity lies in the piecemeal *transformation* of information. Through ingenious and piecemeal acts of artifice – in the field, the lab, the drawing board, the paper –, scientists seek to establish viable *chains* of reference that connect copiously with distant, obscure and hard to understand matters – regarding, say, the firings of neurons – by transforming them at each stage into increasingly manipulable versions of themselves (they remain 'themselves' precisely to the extent that each link of the chain is connected):

[Scientific] knowledge, it is not a direct grasp of the plain and the visible against all beliefs in authority, but an extraordinary daring, complex and intricate confidence in chains of nested transformations of documents that, through many different types of proofs, lead away toward new types of visions which force us to break away from the intuitions and prejudices of common sense. (Latour 2010a 122)

According to Latour, a symmetrical paradox holds for religion. Traditionally it has been assumed of Christianity at least (the point is moot, and we'll come to it presently) that faith is directed at the obscure and the unattainable: mysterious and subjective stuff that lies beyond the transparent objectivity of science. But just as the illusion of representational transparency in science results from bracketing off the long 'distances' traversed by scientists' fragile chains of reference (see, for example, Latour 1999), so, for Latour, the notion of transcendence in religion (the distance faith is supposed to 'leap' across) ignores the *intimacy* religious expression is meant to elicit. For in religion, as in love, the message is the medium. 'I love you' is not a statement that conveys information – if it were, Latour urges, lovers would hardly ask each other to repeat it, time and again. It is a statement that, while utterly banal in itself, strives to bring the utterer and his interlocutor closer. It is just this delicate operation, in which subtle sensibilities as to tone and timing are so vital, that religious speech seeks to enact. Latour chooses as his example the devotional role of iconography in Catholic art, but the point could be made as well with reference to conversion, the Eucharist or the Life and Passion of Jesus: each of these 'speech-acts' conspires to efface its own propositional force so as to render present, in the act, the intimacy of relation itself:

[I]f, when hearing about religion, you direct your attention to the far away, the above, the supernatural, the infinite, the distant, the transcendent, the mysterious, the misty, the sublime, the eternal, chances are that you have not even begun to be sensitive to what religious talk tries to involve you in. . . . In the same way as . . . love sentences should transform the listeners in being close and present or else are void, the ways of talking religion should bring the listener, and also the speaker, to the same closeness and to the same renewed sense of presence – or else they are worse than meaningless. (Latour 2010a: 105)

What makes Latour's argument so bold and provocative is partly its combative sense of revisionism. You thought science was about getting the facts straight, but let me tell you why it is not! You thought religion was grasping for the unknowable, but let me show you how it is about embracing what is most present! As well as posing analytical challenges, Latour's arguments entail indictments of false consciousness, too. Certainly, it is clear that Latour's revisionism is theoretically motivated by his trenchant critique of representationism (as well as his own convictions as a practising Catholic, no doubt – Tresch 2013; cf. Skirbekk 2015), advanced in favour of the unmistakably *relational* analytics of his concept of mediation, which lies at the root of his symmetrical argument on science and religion. There is no ontological discontinuity between word and world – science is no mirror of nature – and, remarkably, there is no such discontinuity between (human) word and God either. Statements about the world and about God should not be taken as attempts to determine how things stand 'out there' or 'up there', for each statement creatively re-casts our relationship with what it is we talk about: in science words bring the most obscure facets of the world closer by transforming them, and in religion words, paintings and other forms of religious expression erase the very possibility of distance between word and God by rendering Him present in the utterance.

In fact, there is an important logical complementarity between Latour's respective points about science and religion. For his point that words can perform ontological work by transforming what they are commonly deemed to represent (as in science) presupposes that the representational capacity of words may be effaced (as in religious expression). Science, in this sense, emerges as a peculiarly pious exercise. A kind of marathon-run of religious intimacy, the 'circulating references' (Latour 1999) that bind the world and science together are 'transformations' only if premised on ontological proximity – the same proximity that religious speech is supposed to elicit. God as Actor Network, or something like it.

From the point of view of our argument about the fundamentally reflexive orientation of the ontological turn, however, Latour's approach presents us with a basic paradox. On the one hand, his revisionist account of religion (as well as of science) is entirely consistent with the kinds of analytical procedures we have been exploring throughout this book, and not least with the emphasis on the ontological transformations that the central concept of the relation entails, as we saw in relation to Viveiros de Castro in Chapter 4. Latour's analysis, in other words, seems to fall squarely on the trajectory of thinking that we have been examining in the previous chapters, proving its analytical power by extending it to the analysis of Christian faith (and of course secular science).

On the other hand, Latour's analysis also illustrates the broader point we made about his work in Chapter 1, namely its lack of that peculiar form of ethnographic reflexivity we associate with anthropology, which the ontological turn seeks to radicalize. For the point is that Christian faith *does*, as a matter of fact (though by no means always – e.g. see Scott 2005; Luhrmann 2012), centre on questions of transcendence in a host of ways. Indeed, in much of the ethnographic literature on the practice of different forms of Christianity, some of which we have cited earlier, the question of transcendence is treated as near-definitive of this form of religious experience, distinguishing it from, say, animism (Keane 2007), polytheism (Holbraad 2012; cf. Assman 2008) and other forms of worship that anthropologists may study. Christians often *do* think of God as 'infinite, distant, transcendent, mysterious', notwithstanding Latour's afore-cited admonition to the contrary. Indeed, is it not precisely because this is so that Latour's argument needs to be so combative in its attempt to establish the opposite by way, as it were, of *conversion* from transcendence to immanence? So, if one accepts – takes seriously – the significance of transcendence as an ethnographic premise,[10] then the flow of Latour's relational analysis runs directly counter to that of the ontological turn. His relational framework is affirmed at the *expense* of the ethnography of Christianity, modifying it in a revisionist manner, rather than being itself revised by it (not unlike, perhaps, what Mosko was earlier found to do with respect to Christian conversion).

This, then, is the point at which the prospect of a post-relational extension of the ontological turn once again presents itself. Let us take it as a given – rather than a sin, as Latour seems to have it – that the ontological distances that transcendence seems to imply are a major concern for at least some versions of Christian worship. To put it in emblematic (and entirely indicative) terms, after the Fall, humanity is estranged from God: we no longer enjoy Him in immanence, but rather must have faith in Him as a transcendental guarantor of Creation (cf. Ricoeur 1998; Engelke 2007: 12–13). Although He gave us his Son and the Sacra as media for our return to Him, this is only testimony to the fact that the distance is there – i.e. a predicament with which the faithful (as well as theologians) must contend (at least in some versions of Christianity), rather than merely an analytical category mistake, as Latour's analysis implies. So what would happen if, rather than painting Christianity over with a relational brush, we instead treated it as an opportunity to keep asking new questions about what a relation might be? In particular, if Christian

[10] Note that for this argument to stand it is not necessary to settle, whether theologically or ethnographically, the vexed question of how far and in what way, if at all, Christianity can be 'defined' as a religion of transcendence (e.g. see Cannell 2006; Lilla 2008; Robbins 2012; cf. Scott 2005). Much as was the case with our earlier discussion of conversion as rupture, the present argument runs merely on the premise that *some* of the time, in *some* ethnographic instances, on *certain* theological renditions, and in *some* senses, transcendence *does* feature.

concerns with transcendence seem naïve from Latour's relational point of view, then should he not be 'taking the natives seriously' just like one would (or ought to) when studying, say, Amazonian animism?

Post-relational questions, then, arise: How might Latour's apparently asphyctic universe of relations (the myriad mediations of the actor network) accommodate the kind of *difficulty* with relations that is implicit in Christian concerns with transcendence? How might the seemingly non-relational idea of transcendence – a kind of ontological rupture, after all – be conceptualized without, crucially, falling back into the mysterious antinomies of the modern project of purification? How, in other words, might the relation itself be extended (transformed, redefined) so as to *include* its own putative opposite? Note how consistent these post-relational questions are with the basic orientation of the ontological turn. As we have seen throughout this book, the ontological turn was itself 'invented', in Wagner's terms, by altering the conceptual coordinates of previous anthropological 'conventions', and not least the conventional idea that anthropology is in the business of representation. One of the prime results, as we have seen, has been an abiding and conceptually sophisticated emphasis on the role of relational transformation in ontologically informed analyses – an idea that is also central to Latour's approach. The post-relational challenge that we are raising, as we also explained in the introduction to this chapter, consists in taking just this idea – the relation – as a new conceptual baseline, ripe for *further transformation* when confronted with ethnographic material that seems to cut against it, such as the apparently non-relational characteristics of Christian faith. The relational infrastructure of the ontological turn, in other words, is here treated as the convention against which relationality itself must be invented to take Christianity seriously. Once again, then, this is the ontological turn operating on itself, shifting its own conceptual coordinates.

As we also stressed in the introduction, the present analysis (like the one of the preceding section) does not pretend to have accomplished this task fully. Our objectives are more narrow and more tentative: we wish to use the example of Christian transcendence, and the challenge it presents to the relational framework of Latour (because he shares it with the ontological turn) as an opportunity, first of all, to show that a post-relational 'move' is indeed possible – indeed, it is required – and, furthermore, to illustrate what such a move could look like by sketching out one way it could be made to work.

Faith in Motion? A Speculative Illustration

As a first step towards imagining how Latour's relational framework could be transformed, we may start by noting the basic antinomy that his argument for it presents – a meta-purification of sorts. On one shore we have the armoury of the Network, based on the logical priority of transformative relations: relations precede entities, in the sense that the latter are the *products* of ontological transformation. On the opposite shore we have representationism, based on the logical priority of self-identical entities, separated from each other by extensive gaps of what we might call, if not transcendence, then simply negativity: either this, or that, with a logical gap in between – negative space. Setting Latour's argument up in this way, the question becomes how the concepts of relation and negation might be brought under a single analytical scheme – how they may be 'hybridized', to use the vocabulary Latour himself favoured in earlier parts of his career (e.g. 1988; 1993). What makes such a project an extension of the relational framework is the fact that rather than negating either concept (as Latour does with negation itself), our task here is to bring them together by *transforming* both of them (adopting, that is, the same approach we took with respect to the distinction between relationism and individualism in the previous example about Christian conversion). That is, since 'ordinary' concepts of relation (*sensu* Latour et al.) and negation (the either/or logic of representation) are antinomous, the task is to redefine them in an 'extraordinary' way that would overcome the antinomy.

After the Relation

A clue as to how this might be done lies in Latour's own contrast between the movement of the relational transformations that ('true') science and religion involve with what he calls the 'freeze framing' stasis of representationism:

[I]n the cases of both science and religion, freeze framing, isolating a mediator out of its chains, out of its series, instantly forbids the meaning to be carried in truth. Truth is not to be found in correspondence – either between the word and the world in the case of science, or between the original [e.g. a transcendent God] and the copy [e.g. an image of Him, or a statement about Him] in the case of religion – but in taking up again the task of continuing the flow, of elongating the cascade of mediations one step further. (Latour 2010a: 122–23)

This question of 'flow', we would like to suggest, goes to the heart of the matter regarding the relationship between relation and negation. Inasmuch as Latour's network of relations is constituted by transformations, it follows that the priority of differential relations over self-identical entities in Latour's relationism should be supplemented by the logical priority of motion over rest: the 'cascade of mediations' which lie at the heart of Latour's analysis is not only a relational field but also a *motile* one (see also Holbraad 2012). This is important because, as we suggest, it renders explicit a sense in which *negativity* can be conceived as constitutive of transformative relations, though this sense is appropriately different from that of negation as it is ordinarily construed in a representationist framework. In the programmatic, sketchy spirit of our illustration, and with apologies to the logicians (as well as to those who dislike their formalism), let us render the thought in purely formal terms.

Consider an ontological transformation from A to B. As a transformation of A, B is not just related to A (viz. 'A – B') but is also the product of it, i.e. the two are related by a vector (i.e. 'A → B'). Now we may ask: how might this vectoral relation be distinguished from another (e.g. how might the vector running from A to B be distinguished from another one running from A to C)? The question seems hard to answer

because questions of 'distinction' call to mind ordinary senses of negation and identity (e.g. A → B 'is not' A → C), which are banished from
the conceptual language of relationism, in which such gaps of negativity
are anathema. Rather, to stage a postrelational move, what is needed is a
conceptualization that would give the same results (distinguishing, say,
A → B from A → C) in a relational manner, without appeal to 'either/or'
negations. Or, to put it conversely, the question is whether one can retain
criteria of distinction (of 'negativity' in a peculiarly minimal sense) in
a purely 'positive' logical universe, where the only available connective
is 'and' – the relational connective (see also Viveiros de Castro 2003;
Pedersen 2011: 206–10).

The answer might be this: in such a purely 'positive' universe distinctions must be made in positive terms, i.e. not as a matter of this 'or' that, but
as a matter of this 'and' that. But this simple reversal – which is really just
a restatement of the premise of relationism – has consequences that are as
important as they are counter-intuitive. To see this it is useful to spell out
the shift from thinking about distinctions in terms of 'and' (i.e. relations
of conjunction) rather than 'or' (i.e. relations of disjunction): instead of
expressing the distinction between our vectors A → B and A → C by writing
'(A → B) or (A → C)' we would now do so by writing '(A → B) and (A → C)'.
In other words, in the relentlessly 'positive' universe of relationism, a given
transformation (or 'vector') can be distinguished from another if and only
if there is a third one that conjoins them. So rather than distinguishing a
transformation in terms of what it is not, we now distinguish it in terms
of what it becomes, where 'becoming' is understood as the transformation that occurs when transformations are brought together in a positive
act of conjunction. Somewhat profoundly, then, the act of distinction is
itself a transformation: to distinguish things is to change each of them by
bringing them together (see also Strathern 2011). Or to put it in twentieth
century pop science terms, to know something is to change it. Or even, to
return to a central idea of our exposition of Strathernian comparisons, to
know something is to *add* something to it.

Note, however, that while these 'conjunctive distinctions', as we may call them, are themselves transformative relations just like the transformative relations that they combine, there is nevertheless an important logical asymmetry at play here. For, while these distinctions encompass the transformations that they distinguish, the opposite is not the case. For example, 'A → B' and 'A → C' are parts of '(A → B) and (A → C)' but the latter is not a part of each of them. The point can be put in the Aristotelian language of potentiality and actuality: 'A –> B' and 'A → C' have the *potential* to transform into the third transformation that conjoins them together, namely '(A → B) and (A → C)', and the latter *actualizes* that potential. So, simply, distinctions (which is to say, these newly conceptualized conjunctive distinctions) actualize potentialities, to recall also Viveiros de Castro's Deleuzian account of Amerindian myth and perspectivism, which we presented in Chapter 4. In fact this is precisely the advantage of seeing relations in motile terms, as vectors. For the asymmetry introduced by adding an arrow to the end of a line, as it were, to make a vector out of a relation, is constitutive to the logic of ontological transformation, which plays out as a cumulative movement from relatively simple potentialities to relatively complex actualizations.

So the crux of our argument is that this asymmetry of movement – nothing other than its *direction* – provides an opening for redefining (transforming, distinguishing) negation in relational terms. Return to the formal example. Representationally speaking, A → B is distinguished from A → C because A → B 'is not' A → C. Relationally, they are distinguished because they can transform each other by way of conjunction, so as to produce a further transformation, '(A → B) and (A → C)'. In the former case negation refers straightforwardly to the external and extensive differences that distinguish A → B from A → C as self-identical units. But in the latter case a kind of negation enters the picture as well, not in the logical form of 'not', but as a kind of 'not yet'– the *positive* 'not' of potentiality – the 'could be', as it were. For now the distinction between A → B and A → C is recast as a matter of what each of them can become (viz. '(A → B) and (A → C)'); as a matter, in other words, of internal and intensive self-difference, *projected at one step remove* as a *potential* transformation. A paradoxical expression seems fair: in the motile network of relationism everything is what it is because of what it is not yet (though one would be tempted to hyphenate this as 'is-not-yet', to show that what is at issue here is not an ordinary privative negation, but rather the positive negation of 'potential').

Of course, all these abstractions only offer a cursory sketch of the 'motile logic' of one possible rendition of 'post-relationism' (though see Holbraad 2012). What they do suggest, however, is that Latourian concepts may well admit further, postrelational elaboration so as to allow an engagement with the practice of Christian worship that is less revisionary than Latour's own, and particularly one that is better equipped to deal with the notion of transcendence that is so often one of its defining characteristics. Recall Latour's central claim, namely that religious expressions in Christianity are meant to generate our proximate relation to God by conspiring to cancel the possibility of representation. The problem, as we have suggested, is that, for some Christians at least, human beings' relation to God is *not* initially proximate, as indicated, for example, by the common Christian conviction that after the Fall the human predicament is one of estrangement (in which case, far from an illusion or a representationalist misunderstanding, the transcendence of God is a basic cosmological premise). Unlike Latour's notion of intimacy (analytically cast as the 'relation'), the idea of motility (the 'vector') is able to render transcendence as an irreducible dimension of human beings' relationship with God. On this analysis, transcendence is not to be understood as a mysterious alterity, characteristic of representationist dualisms (the incommensurability of humans vs. God). Rather it should be taken as a logical constituent of a particular and new *kind* of relationship (cf. Corsín-Jimenez 2007), namely that of transformation, properly construed as a motion that relates terms (humans and God) always at one logical step removed, as a potentiality is related to its own actualization. Human beings are what they are because of what they are-not-yet, the 'yet' here being that of salvation, perhaps – the *hope* of the immanence of God, or of communion.

Indeed, such an analytical rendition would serve also to reflect the irreducibly *hierarchical* relationship between humans and a God that they deem to be transcendent. The transformation from relatively simple potentials to relatively complex actualizations (complex because they are achieved by way of the additive, 'positive' relation of conjunction) is asymmetrical, as we saw, and can in fact be conceived as a form of ascent: vectors conjoined with vectors, conjoined with further vectors, and so on. Sticking to the schematic, purely abstract image that this motile language paints, God Himself might even be conceived as the plenum of full actualization: what happens – perhaps never – when all potential transformations that could ever take place, do so.

By taking God's transcendence – His constitutive distantiation from humanity – seriously, in this last section we have sought to show what a quintessentially post-relational analytical move might involve, providing a thoroughly experimental sketch of what a reconceptualization of the relation could look like. By rendering Latour's networks of relations God-like, if you like, (as opposed to the other way around), his analytic scheme is changed from a horizontal – or, as it is sometimes said, 'flat' – chain of mediations making myriad differences to each other into an intensive hierarchy of more or less transcendent positions. Thus the concept of the 'relation' ceases to be what it was before: by suggesting how a new manner of conceptualizing relations can be motivated by the 'non-relational' character of Christian faith, the relation effectively becomes a vector – just as 'relational', but apparently agile enough conceptually to deal with the challenge of 'negation' that Christian transcendence seems to imply. Crucially, this is neither a backward move nor a compromise with respect to the methodological and theoretical orientation that Latour's analysis, as well as the ontological turn, performs. For the task of developing postrelational concepts that make sense of apparently non-relational materials cannot be a matter of a retreat to modernist naiveté, precisely because, as Latour has taught us, the modern constitution, along with representationalist analyses that it prescribes, cannot make sense of itself.

Nevertheless, perceptive readers may be experiencing an uncomfortable sense of deja vu at this point (e.g. Heywood 2012), noting the close similarity between the illustration of post-relationism apropos Christianity, and the analysis of the role of motility in the cosmology and ritual of Afro-Cuban divination, presented in the previous chapter. In the two cases, it would seem, a near-identical move from relations to vectors is made to deal with questions of cosmological transcendence. This similarity is uncomfortable because, as argued explicitly in the analysis of the role of *aché* in Ifá divination, the problems that the notion of transcendence presents in the Afro-Cuban case and in Christianity are quite different – a matter of relative ontic distances (in the former) as opposed to a question of ontological rupture (in the latter) (see also Holbraad 2012: 109–30; Holbraad in press a; cf. Holbraad & Willerslev 2007). So dealing with both of them by appeal to the notion of movement, particularly with the help of the vectoral concepts of distance and direction that it entails, raises as many analytical questions as it answers. How can the idea of directed motion be used in one case to account for the kind of cosmological continuity that allows divinities and humans to relate during divination, while in the other to account for a sense of basic discontinuity that renders communion between humans and God a miraculous traversal? Is the notion of movement enough to account for these differences, or might the respective conceptualizations need to be elaborated further, perhaps giving the movements in question different shapes in either case, with reference to different conceptual coordinates (e.g. Holbraad in press a)? Are we even talking about the same *kind* of movement in either case? For example, are the distances involved of the same order? And might we not think of the direction of motion differently in each case, e.g. from distance to proximity on the case of divination, as opposed to from the present to the not-yet in Christianity?

Still, the deliberately programmatic character of our present argument allows us to leave these questions hanging, since what is of interest here is precisely the *problem* that these heuristically generated 'controlled failures' raise (see also Introduction). In particular, such demands for further analytical precision serve to illustrate concretely, albeit programmatically, the kinds of questions that a post-relational move on the ontological turn can open up. Our speculative illustration, after all, is not intended to settle anthropologically the grand theological matter of Christian transcendence – as we explained, to do so would involve at the very least a full engagement with the contingencies of particular ethnographic materials. Rather its purpose has been to illustrate the kinds of difference pursuing such a line of thinking can make, not least to the way we conceive of the ontological turn. Hence, the fact that transcendence in the ethnography of Ifá divination and in (any particular ethnography of) Christianity may require our analyses to take the post-relational move in further, indeed *different* directions, beyond just thinking of both of them in motile terms, is grist to the post-relational mill: the more of just this kind of conceptual self-differentiation the better, from the point of view of the ontological turn!

How to Keep Turning

Still, this observation about similarities between the analyses of otherwise divergent ethnographic situations serves to illustrate also a more general and critical point that could be made regarding the kinds of analyses the ontological turn has thus far tended to deliver. For looking from the middle distance at the analytical frameworks that we have been examining throughout this book, one might observe that, different as they are in their substance, they also share some basic continuities. Relationism, as we have discussed in this chapter, is one of them. The idea of transformation, and particularly self-transformation, is another (the notion of motility we have elaborated here and in the previous chapter relies on it). In general, one might say that, notwithstanding the sophisticated differences between the various analyses this book has sought to unpack (Wagner, Strathern and Viveiros de Castro, for example, are *hardly* saying the same thing, as we have shown in detail from chapter to chapter), there is nevertheless a sense that they all share a basic conceptual language, displaying a similar analytical 'feel' or 'aesthetic'. The similarity between our suggested analyses of transcendence in Ifá and in Christianity, then, may be symptomatic of a broader tendency towards analytical affinity in the works of contributors to the ontological turn (see also Holbraad in press b).[11]

Were this point to be fair (and this remains a big 'if'), it would indeed present an important challenge to the ontological turn. For the promise that the ontological turn holds up for anthropology turns above all on the idea that the possibilities for conceptual creativity that lie at its core are

[11] One may ask to what this apparent lack of parsimony might be owed. Should we take the similarities between ontological turn-style analyses as an indication that the ethnographic situations that precipitate them are not, after all, as different from each other as we might imagine? Are there perhaps basic continuities between otherwise vastly different peoples, which are then reflected in the analytical continuities between their conceptualizations by anthropologists? Wagner's seemingly sweeping contention that tribal, peasant and religious people in general take invention to be the basic human task, as we saw in Chapter 2, may serve as an example of this way solving the problem. Or might we say instead that responsibility for the problem lies with ourselves as analysts rather than with the ethnographic world, and admit that what we are confronting here is some order of failure of conceptual imagination on our own part? This would be dismal, since it would effectively amount to the ontological turn admitting that its prime task, that of creating the conditions for ethnographic differences to make a difference, had effectively failed. So we might prefer to adopt a third view, saying that the problem lies neither with a lack of alterity in the world, nor with a lack of imagination on the part of the anthropologist in her attempt to derive concepts from it, but rather in the relationship the ontological turn posits between these two levels – ethnographic variety and its conceptualization. Maybe, according to this view, the concepts that emerge out of our ethnographic engagements tend to seem similar because they are somehow tainted or otherwise influenced by the very *manner* or *method* by which they are derived. Maybe the ontological turn itself is just too 'noisy' or powerful, generating concepts from here, there and everywhere, but somehow, and perhaps only to a certain extent, in its *own image*.

generated out of the contingencies of ethnography, in all of their marvel-
lous variety. If the ontological turn seeks constitutively to keep open the
horizons of anthropological thinking, then its ethnographic warrant for
doing so is precisely the variation that ethnographic research is able to
present, which is in principle inexhaustible, and which the kind of analysis the ontological turn envisages transmutes into new forms of thought.
Thus understood, anthropological thinking depends on the *parsimony* it
is able to establish between the variety of ethnographic experience, so to
speak, and the variation of conceptual effects that it seeks to extract from
this ethnographic material, by letting the anthropological analysis 'run'
through it as a heuristic 'machine'. Crudely put, one might expect the
figurative (though demonstrable) differences between, say, Amazonia,
Melanesia, Cuba, and Mongolia, let alone the United States, Denmark
and the United Kingdom, or spirit possession, day labouring, aquatic
science, and matrifocal kinship, to be reflected in the analytical schemes
each of these objects of ethnographic study can generate. In this conceptual exuberance would lie the delight of anthropology.

Still, the challenge that the sense of similarity in the analytical outputs
of the ontological turn poses – its apparent lack of parsimony faced with
the sheer variety of ethnography – also presents a positive opportunity
for the ontological turn, marking out an horizon for where it might go –
or 'turn' – next. For the point is that what the ontological turn seeks for
itself, constitutively, is to *keep turning*. And turning in its most thoroughgoing orientation, as we have seen throughout this book, is fundamentally
a reflexive exercise – it is above all of a matter of a turn turning on itself.
The promise that the ontological turn presents, one might say, is a machine
for thinking in perpetual motion (Holbraad 2012) – an excessive motion,
ever capable of setting the conditions of possibility for its own undoing.
So the challenge for the ontological turn is, precisely, how to extend this
motion, as a permanent condition of possibility for thinking (until that is,
this principle itself is reflexively transformed into something different –
the ontological turn must aspire, positively, never to become permanent!).

Our argument in favour of a 'post-relational' move at the present juncture, and the tentative and incomplete manner in which we have sought deliberately to illustrate it in this chapter, serves to emphasize exactly where one of the most pressing tasks for the ontological turn may lie at the present moment. In principle, we emphasize, the task will always be different at different times – such is the very logic of motile transformation: in a universe in which thinking is defined as a form of motion ('turning'), what is a problem for thought at one moment is replaced by another one at another time. But for now, at the present juncture of the ontological turn, as we have sought to articulate it in this chapter, what is most needed to keep the turn going is to 'take seriously' ethnographic phenomena that cut against the basic axes of continuity that unite otherwise divergent approaches contained within it. We have seen how Christianity might serve as one front of and for such transformations, deliberately jeopardizing and thus *distorting* the concept of the relation. But other quintessentially 'modern' concerns suggest themselves (see also Candea & Alcayna-Stevens 2012): truth by correspondence, arbitrary signs, referential meanings, causal chains, subject/ predicate semantics, mind/ body distinctions, individuals and social contracts, free will and personal responsibility and so on – all these conceptions, in their diverse ethnographic manifestations, should no longer be discredited as analytical problems waiting to be debunked. Rather, they are best treated as ethnographic objects that beg for original anthropological analysis, by which we mean the search for new conceptualizations that may make a difference to how we think, and not least of how we think from within the ontological turn which, perforce, these experimentations would extend, transform, distort, disfigure or even obliterate. That is the risk. But also the pay-off.

Conclusion

Thinking back on our time as graduate students at British and Danish anthropology departments around the turn of the millennium, both the interest in phenomenology which was so prevalent in the early 1990s, and the subsequent turn to ontology that began to gain traction in Europe a decade later, were generational responses towards the defeatism that marred the discipline around the fall of the Berlin Wall. For example, Pedersen still recalls the sense of gloom, apathy and disappointment when his first-year teachers at Aarhus University, unable to hide their nostalgia for the grand scholarly ambitions of old anthropological masters such as Lévi-Strauss and Bateson, solemnly declared that ethnographic texts were all about the authors themselves, and that our role as anthropologists was to debunk all such hegemonic representations of exotic Others. A necessary break with grand narratives and positivist and colonial ghosts, to be sure, but not very inspiring and imaginative for a young student.

For the same reason it is hardly surprising, in retrospect, that phenomenology and its call for studying 'things as they are' (Jackson 1996) was greeted with open arms by so many prominent anthropologists (e.g. Desjarlais 1992; Csordas 1994; Stoller 1995; Kapferer 1991; Ingold 2000), including a number of people from the generation completing their doctoral studies in the United Kingdom and Denmark in this period (see e.g. Pedersen 2003; Dalsgaard 2004; Willerslev 2004; Vigh 2006).

With its focus on subjective experience via a Husserlian bracketing of the external world, and therefore also of the desire for detached language and objective knowledge, it became possible to do ethnographic research that did not terminate in postmodernist solipsism and purely negative deconstruction,[1] and instead used the fieldworker's social relations with his or her interlocutors as a legitimate and indeed superior source of intersubjective insight (Jackson 1998; Desjarlais & Throop 2011). At the same time, phenomenology's scepticism towards concepts, theories and abstractions of all sorts tied in well with the practice-theoretical church to which virtually all anthropologists had converted with the critique of structuralism, dogmatic Marxism and other grand theories in the late 1970s and early 1980s (Bourdieu 1977; Ortner 1984). Phenomenology in its various anthropological guises (from existentialist humanists, through embodiment theorists, to Heideggerian romantics) allowed for the rebirth of the classic Malinowskian aspiration to grasp the native's point of view, but in a philosophically beefed-up, reflexive form.

---

[1] By 'deconstruction' we refer to forms of cultural critique that aim to debunk different kinds of hegemonic reifications, essentialisms and, indeed, metaphysics and ontologies. We thus use the term advisedly, because anthropologists have used it in this more vague and generic way since the 1980s. Take, as an example of this notion of (de)construction, the following observation (which, incidentally, prefigures some of the concerns of the ontological turn) by Michael Taussig from his influential book *Mimesis and Alterity*: 'With good reason postmodernism has relentlessly instructed us that reality is artifice yet, so it seems to me, not enough surprise has been expressed as to how we nevertheless get on with living, pretending ... that we live facts, not fictions ... When it was enthusiastically pointed out ... that race or gender or nation ... were so many social constructions, inventions, and representations, a window was opened, an invitation to begin the critical project of analysis and cultural construction was offered ... a preamble to investigation [which], by and large, [has] been converted instead into a conclusion – e.g. "sex is a social construction", "race is a social construction", "the nation is an invention", and so forth ... The brilliance of the pronouncement was blinding. Nobody was asking what's the next step? What do we do with this old insight? If life is constructed, how come it appears so immutable? How come culture appears so natural? If things coarse and subtle are constructed, then surely they can be reconstructed as well?' (1993: xv-xvi).

Conclusion

We recall how the first appearance of the debate about ontology in the late 1990s for us represented an odd but exciting displacement of the phenomenological project, a sort of parallel theoretical universe that *also* allowed us to study 'things as they are', but in a subtly different and somehow more radical way. Like the phenomenologist, the ontologically inclined anthropologist is sceptical towards the once hegemonic anthropological and sociological orthodoxy, which dictates that human social and cultural life takes the form of 'representations', and that such 'signs', 'discourses' and so forth can be deciphered and interpreted by means of models and methods imported from linguistics and the philosophy of language. Yet, it is also here, in their shared non-representationalism and critique of logocentrism that the two orientations, phenomenology and the turn to ontology, depart from one another. For whereas phenomenologists tend to look for 'things as they are' in a purported 'everyday experience' held to exist beyond, or even prior to, concepts and language, for the proponents of the ontological turn, conceptualization, abstraction and theorization are intrinsic not just to the project of anthropology itself, but also to ethnographic reality and the lives of our interlocutors. 'Pure practice', after all, 'exists only in theory; any theory is a mode if practice' (2003: 13), as Viveiros de Castro put it in one of those sentences that really struck a chord with us at the time (along with his snipe at postmodernist deconstruction of the negative, debunking sort: 'there are already too many things which do not exist', 1998: 469). To us, then, the ontological turn was exciting partly because it seemed to overcome a blind spot shared by postmodernists and phenomenologists alike: namely, their tendency to caricature the significance of concepts and their relations to the world anthropologists are charged with understanding. It was just such an aversion to concepts that ontology 'turned' away from.

Nevertheless, as it gradually became clearer to us as we familiarized ourselves more with the literature and with the history of the discipline at large, the diverse strands of anthropological thinking that seemed to be coming together in the ontological turn were far from new. Instead of representing a radical rupture from earlier analytical approaches, the turn to ontology from the 1990s onwards had deep links with key figures in the history of anthropological theorizing, going back as far as Mauss, Sapir, Hallowell, Radcliffe-Brown and Lévi-Strauss, as well as influential ideas from the discipline's more recent past, in particular in relation to the work of Wagner, Strathern and Viveiros de Castro, but also the writings of Bruno Latour, Donna Haraway, Tim Ingold, Alfred Gell, Bruce Kapferer, Terry Evens and Anna Tsing, among others. Indeed, returning to our student frustrations, this was another major attraction of the ontological turn when we first learned of it during our graduate years: the fact that the scholars associated with it dared to not just mention, but actually use, some of the very authors and concepts from the three classic traditions of anthropology, which had been dismissed so categorically by influential anthropologists in the 1980s.

This is not to say that postmodernism generally and the crisis of representation in particular were not important and indeed necessary. As we have persistently argued in this book, the ontological turn would probably never have happened, and certainly not have taken the super-reflexive form it has, were it not for the decisive role that the crisis of representation played as its theoretical and methodological precursor. Still, the sometimes earth-scorching way in which key figures from the anthropological past were dismissed and denounced was hardly productive or inspirational, let alone particularly scholarly. Which again explains the sense of relief we experienced when we first read Roy Wagner, Marilyn Strathern's and Viveiros de Castro's work. For a change, Lévi-Strauss and Dumont were not being relegated to the rubbish bin of obsolete and reactionary theory; and, for once, Radcliffe-Brown's and Hallowell's ideas were being taken seriously instead of being reduced to targets of ridicule and scorn.

While this aspect of the trans-Atlantic traffic in anthropological ideas and perspectives remains to be documented in full historical detail, we have in this book sought to trace the core theoretical developments and trajectories that eventually coagulated into the ontological turn. Indeed, as we have wanted to show, the ontological turn may be described as the strange theoretical beast that is born when one takes those visible aspects of social life that the structural-functionalists called 'social relations' and treats them as if they were imbued with the same symbolic properties and efficacies as the hidden meanings unearthed by Lévi-Strauss and Foucault, or, in a different way, the patterns of meaning deciphered by Boas and Geertz. After all, is that not what Wagner's theory of obviation and Strathern's relational ethnography boil down to: a sort of hyper-structuralism, on which everything is treated as if it belonged to a single yet endlessly self-differentiating totality? To 'take the ontological turn' is, for analytical and other experimental anthropological purposes, to treat all things *as if* they were subject to perpetual self-differentiations. Doing so means that the concepts by which a given ethnographic encounter is analysed must be in an ontological continuity with this encounter, *as if* everything subject to anthropological study pertained to one myth-like, intensive universe.

But if the ontological turn is nothing but the realization of an old post-structuralist desire to set free theoretical promises dormant within classic structuralist and symbolic anthropology, then why does it bear this ungainly label? Just how, many anthropologists keep asking themselves with varying degrees of shock, incredulity and impatience, can 'ontology', with its heavy freight of philosophical baggage, be of any use to us anthropologists? Our response to this question has been that it is precisely *because* of its philosophical connotations that the term 'ontology' is apt for formulating a new way forward for anthropology, towards what we consider to be its natural position as a 'savage' alternative to philosophy. What we wish to achieve in using the 'philosophical concept' of ontology (as if concepts could be owned by particular disciplines bestowed with a unique right to deploy and define them) is the exact opposite of the transcendental truth-goal of traditional metaphysics. Instead of building philosophical castles, the core aim of the ontological turn in anthropology is relentlessly to challenge, distort and transform all things, concepts and theories pretending to be absolute (including the metaphysical idea, prevalent among posthumanists, that everything exists in a permanent state of 'becoming' or 'emergence'), by strategically exposing them to ethnographically generated challenges and paradoxes that can systematically undermine them.[2] Or, put otherwise, the ultimate goal of the ontological turn is to take the things that people in the field say, do or use so seriously, that they trump all metaphysical claims made by any political, religious or academic authority, including (and this is where things may become both tricky and interesting) the authority that we assume in making *this* very claim.

This, then, is how the ontological turn transforms the 'negative' procedure of critical deconstruction into a 'positive' procedure for (equally critical) construction. On the 'negative' version that left us so frustrated about the direction in which this discipline seemed to be moving during our student years, anthropological reflexivity took the form of a critical debunking of positiv(ist) representations in order to expose the hegemonic (social, cultural, political, and so forth) conditions of their production. Conversely, on the more 'positive', ontological mode of anthropological reflexivity whose intellectual genealogy and modus operandi we have outlined in this book, anthropological reflexivity takes

---

[2] For the same reason, we can only agree with Sarah Green in her observation that 'any anthropological definition is inevitably an intervention ... This includes claims to any inherent (fixed) lack of fixity, of course – for example, claims that everything, including any difference, is constantly in flux (or in a state of continual variation, say), as often appears in scholarship influenced by Deleuze. In saying that all definitions of difference are also interventions, I am not making such a claim (which would require me to define what difference is, in order to establish what is constantly in flux). Rather, my interest is in the political implications of asserting that difference has one meaning rather than another; and I take that to be an ethnographic approach toward the question of meaning, rather than a definitional one. Difference does have meaning; it simply cannot be stated in advance what that meaning might be, or what are its implications' (2014: 2–3; cf. Holbraad 2012; Blaser 2013; Scott 2014).

the form a less sceptical and less negative – but, we argue, *no less criti-cal* – method for generating forms of thinking that are always in per-petual re-construction. Far from representing an anachronistic return to the depoliticized pleasures of grand theory and armchair speculation (though see Willerslev 2011), then, the ontological turn is acutely mind-ful of and respectful towards postcolonial theory and the postmodern collapse of grand narratives, for it treats these as necessary points of departure for its ethnographically derived concepts and interventions.

Post-critical Anthropology

So to take the ontological turn is to open oneself up to (and embrace the uncertainties and potential risks of) the possibility that there might be more than one mode of anthropological critique (and therefore, as we shall explain in the following section, more than one form of anthropological politics also). This might explain why, from the vantage point of prevailing, Marxist-inspired ideas of anthropological critique, the ontological turn can easily look cerebral and disengaged, if not frivolous, sometimes indulgent, or downright irresponsible, and even dangerous. Indeed, how could it be otherwise, given that the general orientation towards theoretical reflexivity, analytical experimentation and conceptual innovation we have sought to delineate in this book often has involved a quite deliberate questioning of some of the most deeply held epistemological-cum-ontological assumptions that underwrite the human(ist) scientific project, including the prevailing notion that proper cultural analysis involves the interpolation of 'critical distances' towards one's object of study by situating it in a 'broader' political-economic and ideological 'context'.

Of course many anthropologists, and often with good reason, rely on this more established concept of critique in their writings. In fact, it may be maintained that the classic anthropological project of 'cultural critique' (Hart 2001) has been reduced to 'sceptical distance' since the emergence of the crisis of representation and postcolonial studies (Said 1978). And to be sure, one can think of many kinds of ethnographic studies which call for a hefty dose of critical distance and anti-essentialist debunking, or run the risk of becoming 'apolitical or, worse, reactionary project[s]' (Kohn 2015: 322). Certainly, we are not suggesting that the ontological turn, including the forms of critique that it makes possible, is always the solution. More modesty and realistically, we suggest that the ontological turn provides a complement to more established forms of social scientific and cultural critique – one that, in certain ethnographic and analytical situations, might nevertheless prove useful, and often also necessary.

As such the ontological turn can be seen as an attempt to experiment with new heuristics of ethnographic description in the hope of forging alternatives that may be complementary to more well-tried forms of cultural critique that seem to have lost some of their traction or even to have 'run out of steam' (Latour 2004c). Certainly, the ontological turn entails an implicit call for a 'critique of critique' (Latour 2004c; Schmidt 2015: 20), if by critique we understand the hegemonic idea of sceptical debunking. Post-ANT scholars have made similar points with respect to Science and Technology Studies (STS), sometimes in connection with explicitly 'ontological' arguments (Jensen & Morita 2012). But perhaps even more so than STS, anthropology – because it is 'concerned with pragmatic, experimental, and humorous encounters with diverse forces that cannot leave the anthropologist unaffected' (Jensen 2013: 329) – is uniquely positioned to pursue this possibility for post-critical intervention.[3] Indeed, as we have sought to argue, it is anthropology's unparalleled ability, but therefore also

---

[3] We use the term 'post-critical' in the sense suggested by Casper Bruun Jensen in his recent comparison of certain alternative strategies for critique within anthropology and STS (2014b; see also Bargués-Pedreny, et al. 2015). Here, critique does not take as its point of departure, or for that matter depends on, 'a general theory or ... a methodology, a normative framework, or a political project' (2014b: 362). Rather, it constitutes 'both the limit and the creative condition of possibility of [postcritical] theories and methods that they cannot provide more

Conclusion

obligation, to play the role of a 'counter-science' in Foucault's sense (1970) by putting *everything* in question, that imbues it with the license and indeed the responsibility to engender and elicit the creative potential of ethnography for redefining what *any given thing* might be. Certainly, as we have demonstrated, this has been a core matter of concern for scholars instrumental to the genesis of what eventually became known as the ontological turn, including not just Viveiros de Castro but also Wagner and Strathern: anthropology's capacity for generating ethnographically derived ontological questions.

So instead of representing a naively uncritical or reactionary de-politicalized stance, the ontological turn can be described as post- or even 'multi'-critical in the sense that it seeks to combine several critical projects in a single mode of anthropological practice.[4] But perhaps it could also be argued, as Ghassan Hage has done in series of recent publications (2012, 2014), that the ontologically informed modes of intervention and critique with which we are concerned here are of a characteristically anthropological kind. As he puts it,

the anthropological tradition has ... provided us with various modes of critical thinking; some are very similar to sociology, in so far as every

than provisional, experimental, and inherently uncertain grounds for research engagements and interventions ... [P]ostcritical anthropologists have [no] need for general methods, abstract principles, or encompassing political projects to go by' (2014b: 362).

[4] Another, more basic sense in which the ontological turn is irreducibly 'critical' has thus to do with its ambition to scrutinize some of the most fundamental conditions of possibility for anthropological thinking and ethnographic knowledge as such. It is in this 'Kantian' sense (cf. Holbraad 2012: 260–5; in press b; Graeber 2015: 23) that the ontological turn is the natural heir to the proud tradition of 'cultural critique' and the reflexive attitude associated with the 'crisis of representation' of the 1980s in particular – indeed, one might venture that, for better or worse (Salmond 2014; Ingold 2014), the ontological turn is the most theoretically reflexive project that anthropology has produced. Certainly, as we have consistently tried to do in this book, it can be argued that the ontological turn 'as a heuristic has emerged, albeit transformed, from the epistemological bonfire of the 1980s. Reborn from its ashes as political critique, radical conceptual innovation, or in some cases both, [it] survives, and indeed has floated to the top of our disciplinary agenda' (Candea 2015).

anthropology necessarily involves a sociology, but there is also a specific critical function that has emerged from the anthropological tradition that is quite unique to it ... [C]ritical sociology uncovers social forces and social relations that are believed to be already having a causal effect on us regardless of whether we are aware of them or not (class relations, gender relations, etc.), critical anthropology invites us to become aware of and to animate certain social forces and potentials that are lying dormant in our midst. In so doing it incites what was not causal to become so! (2012: 288–90)

Again, this is of course not to say that one cannot engage in 'critical sociological thinking' as an anthropologist (in the same way as anthropologists have found inspiration in and drawn on the work of, for instance, social historians, cognitive psychologists, or analytical philosophers). It is only to suggest that, in doing so, one is not intervening in a manner that could perhaps be described as distinctively anthropological.

But how is this concretely done – how to 'animate forces and potentials lying dormant at our midst', in Hage's words? Here, we need to remind ourselves that all ethnographic descriptions, like all cultural translations, necessarily involve a certain element of transformation or even disfiguration. As we discussed in Chapter 4, anthropological interpretations may thus be conceived of as 'controlled equivocations' (Viveiros de Castro 2004), which, far from transparently mapping one discrete social order or cultural whole onto another, depend on more or less deliberate and reflexive 'productive misunderstandings' (Tsing 2005) to perform translations and comparisons, not just between different contexts, realms, and scales, but also within them. This is what distinguishes the ontological turn from other anthropological approaches: not the assumption that it enables one to take people and things 'more seriously' than other analytical approaches, but the ambition, and more importantly the capacity, to take things, if you like, *too* seriously (Pedersen 2013c). This, as we have argued in Chapter 3, goes to the heart of the Strathernian project of 'postplural' comparison: the fact that a given anthropological

Conclusion

description, analysis or comparison involves an intensely abstract conceptual scaling or 'sculpting' that works by eliciting certain dynamics and potentials present 'within things' into intensified versions of these things themselves, not unlike an artist probing and sensing her way through the bundle of forces that the affordances of her materials enable or even compels her to release. Here, comparison becomes the heart of anthropolocal thinking and is an ontological operation: 'The anthropology *of* ontology is anthropology *as* ontology; not the comparison of ontologies, but comparison as ontology' (Holbraad et al. 2014).[5]

As such, the ontological turn may be said to have something unabashedly playful about it that is not far from what Nietzsche set his sights on for philosophy as a 'gay science' (2001), or what Deleuze and Guattari identified as philosophers' key tasks of asking 'idiotic' questions to poke holes in the firmaments of common sense (1994: 62–3). But crucially, this call for a playful, non-sceptical critique is not just an argument about the inherent fun of anthropology and its potential for making controlled equivocations. It is also an argument from necessity. The claim that lies at the heart of this book, and which we have sought to demonstrate in principle and by working it through our exposition of different contributions to this approach, is that anthropology *needs* to take the ontological turn in order to fully to rise the epistemological, ethical and indeed

---

[5] It follows that, rather than intervening 'ironically' or 'sceptically' by standing, or pretending to stand above the world, anthropology, on its post-critical variant, intervenes via 'lateral extensions' or 'prototypes' of it (Maurer 2005; Miyazaki 2013; Corsín-Jimenez 2015). It could even be argued that, instead of making its interventions via the interpolation of 'critical distances', the ontological turn is imbued with the capacity for forging 'critical proximities' that comically render the world into different distortions of itself (Pedersen 2013c; Nielsen & Pedersen 2012). Comedy's critical potential, after all, has well-described by cultural theorists (e.g. Bakhtin 1965; Mbembe 1992; Bergson 1999), and an argument can be made that the time has come to 'go beyond the ontological joke' (J. Schmidt 2015: 22) by systematically investigating the manner in which the powerful logic of the ludic is intimately connected to the mechanisms by which certain kinds of anthropological thinking operate (Pedersen & Nielsen 2015, forthcoming; M. Schmidt 2015; see also Wagner 2001).

political challenges that describing and analysing the contingency of ethnographic exposures present.

The Politics of Ontology

Broadly speaking,[6] we may distinguish between three different ways in which ontology and politics are correlated in the social sciences and the humanities,[7] each associated with particular methodological prescriptions, analytical injunctions and moral visions: (1) the traditional philosophical concept of ontology, in which 'politics' takes the implicit form of an injunction to discover and disseminate a single absolute truth about *how things are*; (2) the sociological critique of this and other so-called 'essentialisms', which, in sceptically debunking all ontological projects to reveal their insidiously political nature, paradoxically runs the risk of ending up affirming and reifying the critical politics of debunking as its own essentialist version of *how things should be* and (3) the anthropological concept of 'the ontological' (not 'ontology' or 'ontologies'!), in which multiple forms of existence enacted in concrete practices reflexively precipitate alternative forms of thinking, so that 'politics' becomes the non-sceptical activity of eliciting a manifold of potentials for *how things could be*.

Accordingly, as we have shown in this book, while the ontological turn in anthropology has made the study of ethnographic difference or 'alterity' one of its trademarks, it is really less interested in differences between things than *within* them: the politics of ontology is the question of how

---

[6] The present section is largely an abridged version of Holbraad, Pedersen & Viveiros de Castro (2014), drawing heavily on our own contributions to this co-authored text, but not on the parts Viveiros de Castro contributed.

[7] Although discussions of 'ontological politics' are relatively new to anthropology (though see Verran 2001; Blaser 2009; Whatmore 2013), STS scholars have used the expression for some time now (Thompson 1996; Mol 1999; Law 2004; Pickering 2011).

Conclusion

persons and things could alter from themselves. The politics of ontology is the question of how persons and things could alter from themselves, precisely because the concept of 'self-difference' is intrinsic to the manner in which anthropological analysis is done. Again, this is the key lesson from Wagner and Strathern, as well as Viveiros de Castro, of which one should not lose sight when articulating what the ontological turn is about, including why it has emphatically *nothing* to do with 'essences', 'cultural differences' or anything of the sort; in fact quite the contrary. As a sort of reverse deconstruction, the ontological turn performs its interventions, not by making the world less real by taking it apart and thus exposing the processes that made it into what it is, but by adding to it – taking it 'too seriously'– and thereby making it 'more' or differently real. Again, critique here happens, not as sceptical acts of debunking, but by performing analytical operations that turn things into what they could be, but still are not.

This, as we have argued in this book, is what the ontological turn is: a technology of description for experimenting with what in Chapter 5 we called the conceptual affordances present in a given body of ethnographic materials. Articulating the study of 'what could be' in this way implies a peculiarly non- or anti-normative stance, which has profoundly political implications in several senses. First, subjunctively to present alternatives to declarations about what 'is' or imperatives about what 'should be' is itself a political act to the degree that it breaks free of the glib relativism of merely reporting on alternative possibilities ('worldviews'), and proceeds to lend the 'otherwise' (Povinelli 2012) full ontological weight as a real alternative.

Secondly, when such conceptual experimentations are precipitated by ethnographic exposures to people whose own lives are, in one way or other, pitted against the reigning hegemonic orders (state, empire and market, in their ever-volatile and so often violent comingling), the politics of ontology can be aligned deliberately with the politics of the peoples who occasion it. We saw one way in which this can work in Chapter 4 in relation to Viveiros de Castro's explicit attempt to connect the ontological turn with a political programme for the decolonization of anthropological thought and its implications for the conceptual self-determination of the peoples we study (see also Blaser 2013; de la Cadena 2015). Indeed, as we suggested in our critical engagement with this way of doing (political) ontology, its experimental gesture may be extended also to the very definition of what 'the political' itself could be in any given ethnographic situation – a postcritical distortion of the concept of politics itself. Certainly, raising such questions would be a further way to render anthropological analysis not merely logically contingent upon, but also morally imbricated with, the political dynamics in which the people anthropologists study are embroiled, including the political stances those people might themselves take.

Finally, the political promise held by ontologically oriented approaches in anthropology and cognate disciplines can be conceived, not just in relation to the degree to which such approaches are in affinity with (or even actively promote) particular political objectives, or with the abiding need for a critique of the state of the world and the forms of thought that underpin it, but also in relation to their capacity to enact a form of politics that is entailed in their very operation. Conceived of in this manner, the ontological turn is not so much a means to externally defined political ends, but, in a certain sense at least, a political end in its own right. Recapitulating, to some extent, standing debates about the political efficacies of intellectual life (e.g. the ambivalent stance of Marxist intelligentsias to communist parties' calls to political militancy in the twentieth century), the question is whether ontologically oriented analyses render political the very form of thinking that they involve, such that 'being political' becomes an immanent property of the mode of anthropological thought itself. If so, then the politics of the ontological turn in anthropology resides not only in the ways in which it may help promote certain futures, but also in the way that it 'figurates' the future (Krøijer 2015) in its very enactment.

The major premise of such an argument might border on a cogito-like form of self-evidence: 'to think is to differ'. Here, a thought that makes no difference to itself is not a thought: thoughts take the form of motions from one 'position' to another, so if no such movement takes place then no thought has taken place either. Note that this is not an ontological credo (e.g. compare with Levi Bryant's recent [2011] 'ontic principle', which is similar, but cast in the philosophical key of meta-physical claim-making). Rather, it is offered as a statement of the logical form of thinking – a phenomenology in Simon Critchley's (2012: 55) sense that is, moreover, self-evident insofar as it instantiates itself in its own utterance. The minor premise, then, would be the (more moot) idea that to differ is itself a political act. This would require us to accept that such non-controversially 'political' notions as power, domination, or authority are relative stances towards the possibility of difference and its control. To put it very directly (crudely, to be sure), domination is a matter of holding the capacity to differ under control – to place limits upon alterity and therefore, ipso facto (viz. by internal implication from the to-think-is-to-differ premise above) upon thought also.

If these two premises were to be accepted, then it follows that a certain kind of politics becomes immanent to the ontological turn. For if it is correct to say that the ontological turn 'turns', precisely, on trans-muting ethnographic exposures reflexively into forms of conceptual creativity and experimentation, then one can also say that it is abidingly oriented towards the production of difference, or 'alterity', as such, with all the epistemological and ethical challenges and responsibilities that follow from making this claim (cf. Green 2014; Hohm 2014; Salmond Forthcoming). Regardless (at this level of analysis) of the political goals to which it may lend itself, anthropology is *ontologically political* inasmuch as its operation presupposes, and is an attempt experimentally to 'do', difference as such: the politics of indefinitely sustaining the possible, the 'could be'. It is an anthropology, then, that is analytically anti-authoritarian, making it its business to generate vantages from which established forms of thinking are put under relentless pressure by alterity itself, and perhaps changed.

Bibliography

Agamben, Giorgio. 2004. *State of Exception*. Chicago: University of Chicago Press.

Alberti, Benjamin. 2014a. 'Archaeology, risk, and the alter-politics of materiality'. Fieldsights – Commentary. *Cultural Anthropology* website, http://culanth.org/fieldsights/476-archaeology-risk-and-the-alter-politics-of-materiality (accessed 28 June 2015).

2014b. 'Designing body-pots in the formative La Candelaria culture, Northwest Argentina', in E. Halam and T. Ingold (eds.), *Making and Growing: Anthropological Studies of Organisms and Artefacts*, pp. 107–26. Farnham: Ashgate.

Alberti, Benjamin and Yvonne Marshall. 2009. 'Animating archaeology: local theories and conceptually open-ended methodologies'. *Cambridge Archaeological Journal* 19 (3): 344–56.

Alberti, Benjamin, Severin Fowles, Martin Holbraad, Yvonne Marshall and Christopher Witmore. 2011. '"Worlds otherwise": archaeology, anthropology, and ontological difference'. *Current Anthropology* 52 (6): 896–912.

Ansell Pearson, Keith. 1999. *Germinal Life. The Difference and Repetition of Deleuze*. London: Routledge.

Appadurai, Arjun. 1986. 'Introduction: commodities and the politics of value', in A. Appadurai (ed.), *The Social Life of Things: Commodities in Cultural Perspective*, pp. 3–63. Cambridge: Cambridge University Press.

1996. *Modernity at Large: Cultural Dimensions of Globalization*. Minneapolis: Minneapolis University Press.

Argyrou, Vassos. 2002. *Anthropology and the Will to Meaning: A Postcolonial Critique*. London: Pluto Press.

Asad, Talal. 1993. *Genealogies of Religion: Discipline and Reasons of Power in Christianity and Islam*. Baltimore: Johns Hopkins University Press.

Assmann, Jan. 2008. *Of God and Gods: Egypt, Israel, and the Rise of Monotheism* Madison: University of Wisconsin Press.

Bibliography

Bakhtin, Mikael. 1965. *Rabelais and His World*. Cambridge, MA: MIT University Press.

Barad, Karen. 2007. *Meeting the Universe Halfway: Quantum Physics and the Entanglement of Matter and Meaning*. Durham, NC: Duke University Press.

Bargués-Pedreny, Pol, Koddenbrock, Kai, Schmidt, Jessica, and Schmidt, Mario (eds.). 2015. Ends of Critique. *Global Dialogues* 10. Duisburg: Käte Hamburger Kolleg / Centre for Global Cooperation Research.

Barker, John. 2003. 'Christian bodies: dialectics of sickness and salvation among the Maisin of Papua New Guinea'. *Journal of Religious History* 27 (3): 272–92.

— 2010. 'The varieties of Melanesian Christian experience: a comment on Mosko's "Partible Penitents"'. *JRAI* 16 (1): 247–9.

Barnard, Alan. 2000. *History and Theory in Anthropology*. Cambridge: Cambridge University Press.

Bateson, Gregory. 1958. *Naven*. 2nd edn. Stanford: Stanford University Press.

Battaglia, Debbora. 1990. *On the Bones of the Serpent: Person, Memory, and Mortality in Sabarl Island Society*. Chicago: Chicago University Press.

Benedict, Ruth. 1934. *Patterns of Culture*. New York: Houghton Mifflin.

Bennett, Tony and John Frow (eds.). 2008. *The SAGE Handbook of Cultural Analysis*. London: Sage.

Bergson, Henri. 1965. *Duration and Simultaneity*. Indianapolis, IN: Bobbs-Merrill.

— 1999. *Laughter: An Essay on the Meaning of the Comic*. Copenhagen: Green Integer.

— 2005. *Matter and Memory*. New York: Zone Books.

Berliner, David, Laurent Legrain and Mattijs van de Por. 2013. 'Bruno Latour and the anthropology of the moderns'. *Social Anthropology* 21 (4): 435–47.

Bertselsen, Bjørn E. 2014 'Maize mill sorcery. Cosmologies of substance, production and accumulation engaged in central Mozambique', in Allen Abramson and Martin Holbraad (eds.), *Framing Cosmologies: The Anthropology of Worlds*, pp. 199–220. Manchester: Manchester University Press.

Bessire, Lucas and David Bond. 2014. 'Ontological anthropology and the deferral of critique'. *American Ethnologist* 41(3): 440–56.

Bhaskar, Roy. 1994. *Plato, Etc.: The Problems of Philosophy and Their Resolution*. New York: Verso.

Bialecki, John, Naomi Haynes and Joel Robbins. 2008. 'The anthropology of Christianity'. *Religion Compass* 2: 1139–58.

Bille, Mikkel. 2013. 'Dealing with dead saints', in D. R. Christensen and R. Willerslev (eds.), *Taming Time, Timing Death. Social Technologies and Ritual*, pp. 137–55. Farnham: Ashgate.

— 2015. 'Hazy worlds: atmospheric ontologies in Denmark'. *Anthropological Theory* 15 (3): 257–74.

Bille, Mikkel, Frida Hastrup, and Tim Flohr Sørensen (eds.). 2010. *An Anthropology of Absence: Materializations of Transcendence and Loss*. New York: Springer.

Binkley, Sam and Jorge Capetillo. 2009. *A Foucault for the 21th Century: Governmentality, Biopolitics and Discipline in the New Millennium*. Cambridge: Cambridge Scholars Publishing.

Bird-David, Nurit. 1999. 'Animism revisited: personhood, environment, and relational epistemology'. *Current Anthropology* 40 (supplement): 67–91.

Blaser, Mario. 2009. 'The threat of the Yrmo: the political ontology of a sustainable hunting program'. *American Anthropologist* 111 (1): 10–20.

——— 2010. *Storytelling Globalization from the Chaco and Beyond*. Durham, NC: Duke University Press.

——— 2013. 'Ontological conflicts and stories of peoples in spite of Europe: toward a conversation on political ontology'. *Current Anthropology* 54(5): 547–68.

——— 2014. 'The political ontology of doing difference and sameness.' Theorizing the Contemporary, *Cultural Anthropology* website. https://culanth.org/fieldsights/474-the-political-ontology-of-doing-difference-and-sameness (accessed 13 January 2014).

Bloch, Maurice. 1977. 'The past and the present in the present'. *Man (N.S.)* 12 (2): 278–92.

——— 1992. *From Prey into Hunter*. Cambridge: Cambridge University Press.

Boas, Franz. 1940. *Race, Language and Culture*. New York: Macmillan.

Boellstorff, Tom. 2016. 'For whom the ontology turns: theorizing the digital real'. *Current Anthropology* 57(4). *Current Anthropology* website, www.journals.uchicago.edu/doi/full/10.1086/687362 (accessed 28 July 2016).

Bogost, Ian. 2012. *Alien Phenomenology, or What It's Like Be a Thing (Posthumanities)*. Minneapolis: University of Minnesota Press.

Bohm, David. 1980. *Wholeness and the Implicate Order*. London: Routledge.

Bonelli, Cristóbal. 2012. 'Ontological disorders: nightmares, psychotropic drugs and evil spirits in southern Chile'. *Anthropological Theory* 12 (4), 407–26.

——— 2015. 'To see that which cannot be seen: ontological differences and public health policies in Southern Chile'. *Journal of the Royal Anthropological Institute (N.S.)* 21: 872–91.

Born, Georgina. 2005. 'On musical mediation: ontology, technology and creativity'. *Twentieth-century Music* 2(01): 7–36.

——— 2010. 'The social and the aesthetic: for a post-Bourdieuian theory of cultural production'. *Cultural Sociology* 4(2): 171–208.

Bourdieu, Pierre. 1977. *Outline of a Theory of Practice*. Cambridge: Cambridge University Press.

Bowker, Geoffrey C. 2009. 'A plea for pleats', in C. B. Jensen and K. Rödje (eds.), *Deleuzian Intersections: Science, Technology, Anthropology*, pp. 123–38. New York: Berghahn.

Bibliography

Brightman, Marc, Vanessa Grotti and Olga Ulturgasheva (eds.). 2014. *Animism in Rainforest and Tundra*. Oxford: Berghahn Books.

Brook, Peter. 1968. *An Empty Space: A Book about the Theatre – Deadly, Holy, Rough, Immediate*. London: Simon and Schuster.

Broz, Ludek 2007. 'Pastoral perspectivism: a view from Altai'. *Inner Asia* 9 (2): 291–310.

Bryant, Levi R. 2011. *The Democracy of Objects*. Ann Arbor, MI: Open Humanities Press

Bryant, Levi R., Nick Srnicek and Graham Harman. 2011. *The Speculative Turn: Continental Materialism and Realism*. Melbourne: re.press

Bubandt, Nils. 2014. *The Empty Seashell. Witchcraft and Doubt on an Indonesian Island*. Ithaca, NY: Cornell University Press.

Bubandt, Nils and Ton Otto (eds.). 2010. *Experiments with Holism*. London: Blackwell.

Buckley, Thomas. 1996. '"The Little History of Pitiful Events": the epistemological and moral contexts of Kroeber's Californian ethnology', in G. Stocking (ed.), *Volksgeist as Method and Ethic. Essays on Boasian Ethnography and the German Anthropological Tradition*, pp. 257–97. Madison, WI: University of Wisconsin Press.

Bumochir, Dulam. 2014. 'Institutionalization of Mongolian shamanism: from primitivism to civilization'. *Asian Ethnicity* 15 (4): 473–91.

Buyandelger, Manduhai. 2013. *Tragic Spirits: Shamanism, Socialism, and the State of Neoliberalism in Mongolia*. Chicago: University of Chicago Press.

Callon, Michel and Bruno Latour. 1981. 'Unscrewing the big Leviathan; or how actors macrostructure reality, and how sociologists help them to do so?', in K. Knorr and A. Cicourel (eds.), *Advances in Social Theory and Methodology*, pp. 277–303. London: Routledge and Kegan Paul.

Candea, Matei. 2010a. '"I fell in love with Carlos the meerkat." Engagement and detachment in human-animal relations'. *American Ethnologist* 37 (2): 241–58.

(ed.) 2010b. *The Social after Gabriel Tarde*. London and New York: Routledge.

2010c. 'For the motion: ontology is just another word for culture. Motion Tabled at the 2008 Meeting of the Group for Debates in Anthropological Theory, University of Manchester', *Critique of Anthropology*, 30 (2), 172–9.

2011a. 'Endo/Exo'. *Common Knowledge* 17 (1): 146–50.

2011b. 'Our division of the universe: making space for the non-political in the anthropology of politics'. *Current Anthropology* 52 (3): 309–34.

2012. 'Different species, one theory: reflections on anthropomorphism and anthropological comparison'. *Cambridge Anthropology* 30 (2): 118–35.

2014. 'The ontology of the political turn'. Fieldsights – Commentary. *Cultural Anthropology* website, http://culanth.org/fieldsights/469-the-ontology-of-the-political-turn (accessed 22 April 2015).

2015 'Going full frontal, or, the elision of lateral comparison in anthropology'. *CRASSH Sawyer Seminar* 25/09/2015. Available at: www.mateicandea.net/?p=962 (accessed 15 February 2016).

Candea, Matei and Lys Alcayna-Stevens. 2012. 'Internal others: ethnographies of naturalism'. *Cambridge Anthropology* 30 (2): 36–47.

Candea, Matei, Joanne Cook, Catherine Trundle and Tom Yarrow (eds.). 2015. *Detachment: Essays on the Limits of Relational Thinking.* Manchester: Manchester University Press.

Cannell, Fanella (ed.). 2006. *The Anthropology of Christianity.* Durham, NC: Duke University Press.

Carneiro da Cunha, Manuela and Eduardo Viveiros de Castro (eds.). 1993. *Amazônia: etnologia e história indígena.* São Paulo: NHII-USP/FAPESP.

Carrier, James (ed.). 1992a. *History and Tradition in Melanesian Anthropology.* Berkeley, CA: University of California Press.

———— 1992b. 'Occidentalism: the world turned upside-down'. *American Ethnologist* 19 (2): 195–212.

———— 1998 'Property and social relations in Melanesian Anthropology', in C. Hann (ed.), *Property Relations: Renewing the Anthropological Tradition*, pp. 85–103. Cambridge: Cambridge University Press.

Carsten, Janet. 2014. 'An inteview with Marilyn Strathern: kinship and career'. *Theory, Culture & Society* 31 (2/3): 263–81.

Chakrabarty, Dipesh. 2000. *Provincializing Europe: Postcolonial Thought and Historical Difference.* Princeton, NJ: Princeton University Press.

Chalmers, David J. 2002. 'On sense and intension', in J. Tomberlin (ed.), *Philosophical Perspectives 16: Language and Mind*, pp. 135–82. Oxford: Blackwell.

Charbonnier, Pierre, Gildas Salmon and Peter Skafish (eds.). 2016. *Comparative Metaphysics: Ontology after Anthropology.* London: Rowman & Littlefield International.

Charlier, Bernard. 2015. *Faces of the Wolf. Managing the Human, Non-human Boundary in Mongolia.* Leiden: Brill.

Chua, Liana. 2009. 'To know or not to know? Practices of knowledge and ignorance among Bidayuhs in an "impurely" Christian world'. *Journal of the Royal Anthropological Institute (N. S.)* 15: 332–438.

Chumley, Lily H. and Nicholas Harkness. 2013. 'Introduction: Qualia'. *Anthropological Theory* 13 (1/2): 3–11.

Clark, Andy. 2008. *Supersizing the Mind. Embodiment, Action, and Cognitive Extension.* Oxford: Oxford University Press.

Clark, Andy and David J. Chalmers. 1998. 'The Extended Mind'. *Analysis* 58: 10–23.

Clastres, Pierre. 1974. "Échange et pouvoir: Philosophie de la Chefferie Indienne." In *La société contre l'État: Recherches d'anthropologie politique*, pp. 25–42. Paris: Minuit.

Clifford, James. 1988. *The Predicament of Culture: Twentieth-Century Ethnography, Literature, and Art.* Cambridge, MA: Harvard University Press.

Bibliography

Clifford, James and George E. Marcus (eds.). 1986. *Writing Culture: the Poetics and Politics of Ethnography*. Berkeley, CA: University of California Press.

Comaroff, Jean and John Comaroff. 1998. 'Occult economies and the violence of abstraction: notes from the south african postcolony'. *American Ethnologist* 26 (3): 279–301.

Conneller, Chantal. 2011. *An Archaeology of Materials*, Cambridge: Cambridge University Press.

Corsín-Jimenez, Alberto. 2007. 'Well-being in anthropological balance: remarks on proportionality as political imagination', in A. Corsín-Jimenez (ed.), *Culture and Well-Being. Anthropological Approaches to Freedom and Political Ethics*, pp. 180–200. London: Pluto Press.

    2013. *An anthropological trompe l'oeil for a common world*. Oxford and New York: Berghahn.

    (ed.). 2014. 'Prototyping cultures: art, science and politics in beta'. Special Issue, *Journal of Cultural Economy* 7 (4).

    2015. Anthropology: a prototype. *Social Anthropology* 23 (3): 348–50.

Corsín-Jiménez, Alberto, and Rane Willerslev. 2007. 'An anthropological concept of the concept: Reversability among the Siberian Yukaghirs'. *Journal of the Royal Anthropological Institute* (n.s.) 13: 527–44.

Course, Magnus. 2010. 'Of words and fog: linguistic relativity and Amerindian ontology'. *Anthropological Theory*, 10: 247–63.

Critchley, Simon. 2012. *Infinitely Demanding: Ethics of Commitment, Politics of Resistance*. London: Verso Books.

Crook, Tony. 2007. *Anthropological Knowledge, Secrecy and Bolivip, Papua New Guinea: Exchanging Skin*. Oxford: Oxford University Press.

Csordas Thomas. 1994. *The Sacred Self: A Cultural Phenomenology of Charismatic Healing*. Berkeley, CA: University of California Press.

da Col, Giovanni. 2007. 'The view from somewhen: events, bodies and the perspective of fortune around Khawa Karpo, a Tibetan sacred mountain in Yunnan province'. *Inner Asia* 9 (2): 215–35.

da Col, Giovanni and David Graeber. 2011. 'Foreword: The return of ethnographic theory'. *HAU: Journal of Ethnographic Theory* 1 (1): vi–xxxv.

Dalsgaard, Anne Line. 2004. *Matters of Life and Longing: Female Sterilisation in Northeast Brazil*. Copenhagen: Museum Tusculanum Press.

Dant, Tim. 2005. *Materiality and Society*. Maldenhead: Open University Press.

Darnell, Regna. 1986. 'Personality and culture: the fate of the Sapirian alternative', in G. Stocking (ed.), *Malinowski, Rivers, Benedict and Others. Essays on Culture and Personality*, pp. 156–83. Madison, WI: University of Wisconsin Press.

De la Cadena, Marisol. 2010. 'Indigenous cosmopolitics: conceptual reflections beyond politics as usual'. *Cultural Anthropology* 25 (2): 334–70.

2014. 'The politics of modern politics meets ethnographies of excess through ontological openings'. Available at: http://culanth.org/fieldsights/463-what-an-ontological-anthropology-might-mean (accessed 15 February 2016).

2015. *Earth Beings: Ecologies of Practice across Andean Worlds.* Durham, NC: Duke University Press.

De la Cadena, Marisol, Marianne E. Lien, Mario Blaser, Casper Bruun Jensen, Tess Lea, Atsuro Morita, Heather Swanson, Gro Ween, Paige West and Margaret Wiener. 2015. 'Anthropology and STS: generative interfaces'. *Hau: Journal of Ethnographic Theory* 5 (1): 437–75.

De Landa, Manuel. 2002. *Intensive Science and Virtual Philosophy.* London: Continuum.

2006. *A New Philosophy of Society: Assemblage Theory and Social Complexity.* London & New York: Continuum.

Deleuze, Gilles. 1993. *The Fold: Leibniz and the Baroque.* Translated and with a foreword by James Conley. London: Athlone.

1994. *Difference and Repetition.* Translated by Paul Patton. London: Athlone.

Deleuze, Gilles and Felix Guattari. 1994. *What is Philosophy?* Translated by Garahm Bell and Hugh Tomlinson. London: Verso.

Demian, Melissa. 2007. 'Canoe, mission boat, freighter: the life history of a Melanesian relationship'. *Paideuma* 53: 89–109.

Descola, Philippe. 1992. 'Societies of nature and the nature of society', in A. Kuper (ed.), *Conceptualising Society*, pp. 107–26. London: Routledge.

1996. 'Constructing natures: symbolic ecology and social practice', in P. Descola and G. Pálsson (eds.), *Nature and Society – Anthropological perspectives*, pp. 82.102. London: Routledge.

2013. *Beyond Nature and Culture.* Translated by Janet Lloyd. Chicago: University of Chicago Press.

2014. 'Modes of being and forms of predication'. *HAU: Journal of Ethnographic Theory* 4 (1): 271–80.

Desjarlais, Robert. 1992. *Body and Emotion: The Aesthetics of Healing in the Nepal Himalayas.* Philadelphia: University of Pennsylvania Press.

Desjarlais, Robert and C. Jason Throop. 2011. 'Phenomenological approaches in anthropology'. *Annual Review of Anthropology* 40: 87–102.

Di Giminiani, Piero. 2013. 'The contested rewe: sacred sites, misunderstandings and ontological pluralism in Mapuche land negotiations'. *Journal of the Royal Anthropological Institute* 19 (3): 527–44.

Dilley, Roy (ed). 1999. *The Problem of Context: Perspectives from Social Anthropology and Elsewhere.* Oxford: Berghahn.

Domínguez Rubio, Fernando. 2016. 'On the discrepancy between objects and things. An ecological approach'. *Journal of Material Culture.* UC San Diego. Available at: http://eprints.cdlib.org/uc/item/09z5n5n2 (accessed 15 February 2016).

Bibliography

Dulley, Iracema. 2015. *Os nomes dos outros: etnografia e diferença em Roy Wagner.* São Paulo: Humanitas, FAPESP.

Dumont, Louis. 1986. *Essays on Individualism: Modern Ideology in Anthropological Perspective.* Chicago: University of Chicago Press.

Durkheim, Émile. 1982. *The Rules of Sociological Method and Selected Texts on Sociology and its Method.* Translated by W. D. Halls. New York: Free Press.

2006. *On Suicide.* Trans. Robin Buss. London: Penguin Classics

Durkheim, Emile, and Marcel Mauss. (1963). *Primitive Classification.* Chicago: University of Chicago Press.

Eliade, Mircea. 1991. *The Myth of Eternal Return, or, Cosmos and History.* Translated by Willard R. Trask. London: Arkana.

Elliot, Alice. 2016. 'The make-up of destiny: Predestination and the labor of hope in a Moroccan emigrant town'. *American Ethnologist* 43(3): 488–499

Ellis, Joe. 2015. 'Assembling contexts. the making of political-economic potentials in shamanic workshop in Ulaanbaatar'. *Inner Asia* 17: 52–76.

Elster, Jon, 1982. 'The case for methodological individualism'. *Theory and Society,* 11: 453–82.

Empson, Rebecca. 2007. 'Separating and containing: people and things in Mongolia', in A. Henare, M. Holbraad and S. Wastell (eds.), *Thinking Through Things. Theorizing Artefacts Ethnographically,* pp. 113–40. London: Routledge.

Engelke, Matthew, 2007. *A Problem of Presence: Beyond Scripture in an African Church.* Berkeley, CA: University of California Press.

(ed.). 2008. 'The objects of evidence'. *Journal of the Royal Anthropological Institute,* 14(S1): S1–S21.

Englund, Harri and Tom Yarrow. 2013. 'The place of theory: rights, networks, and ethnographic comparison'. *Social Analysis* 57 (3): 132–49.

Eriksen, Thomas Hylland. 2007. *Globalization: The Key Concepts.* Oxford: Berg.

Escobar, Arturo. 2007. 'The "Ontological Turn" in social theory: a commentary on "Human Geography without Scale" by Sallie Marston, John Paul Jones II and Keith Woodward'. *Transactions of the Institute of British Geographers* 32 (1): 106–11.

Escobar, Arturo, and Michal Osterweil. 2010. 'Social movements and the politics of the virtual: Deleuzian strategies', in C. Bruun Jensen and K. Rödje (eds.), *Deleuzian Intersections: Science, Technology, Anthropology,* pp. 187–217. New York: Berghahn Books.

Evans-Pritchard, Edward E. 1937. *Witchcraft, Oracles, and Magic Among the Azande.* Oxford: Clarendon Press.

1956. *Nuer Religion.* New York & Oxford: Oxford University Press.

Even, Marie-Dominique. 1988–89. 'Chants de chamanes de mongols'. *Études Mongoles…et sibériennes* 19–20.

Evens, Terence M. S. 1983. 'Mind, logic, and the efficacy of the Nuer incest prohibition'. *Man (N. S.)* 18 (1): 111–33.

2008. *Anthropology as Ethics: Nondualism and the Conduct of Sacrifice*. Oxford & New York: Berghahn Books.

Evens, Terence M. S. and Don Handelman (eds.) 2006. *The Manchester School: Practice and Ethnographic Praxis in Anthropology*. Oxford & New York: Berghahn Books.

Fausto, Carlos. 2007. 'Feasting on people: eating animals and humans in Amazonia'. *Current Anthropology* 48 (4): 518–19.

Ferguson, James. 1994. *The Anti-politics Machine: 'Development,' De-politicization, and Bureaucratic Power in Lesotho*. Minneapolis, MN: University of Minnesota Press.

Fortis, Paulo. 2009. 'The birth of design: a Kuna theory of body and personhood'. *Journal of the Royal Anthropological Institute (N.S)* 16 (3): 480–95.

2010. 'The birth of design: A Kuna theory of body and personhood'. *Journal of the Royal Anthropological Institute* 16: 480–495.

Foster, Jay. 2011. 'Ontologies without metaphysics: Latour, Harman and the philosophy of things'. *Analecta Hermeneutica* 3: 1–26.

Foucault, Michel. 1970. *The Order of Things: An Archaeology of the Human Sciences*. New York: Pantheon Books.

1985. *The History of Sexuality: The Use of Pleasure*. New York: Vintage Books.

Fowles, Severin. 2008. 'The perfect subject: Postcolonial object studies'. Paper at Annual Conference of *Theoretical Archaeology Group*, 24 May, Columbia University, New York.

2010. 'People without things', in M. Bille et al. (eds.), *An Anthropology of Absence: Materializations of Transcendence and Loss*, pp. 23–41. New York: Springer.

Friedman, Jonathan. 2001. 'The iron cage of creativity: an exploration', in J. Liep (ed.), *Locating Cultural Creativity*, pp. 46–61. London: Pluto Press.

2006. 'Comment on Searle's "Social ontology": the reality of the imaginary and the cunning of the non-intentional'. *Anthropological Theory* 6 (1): 70–80.

Fuentes, Ahustín and Eduardo Kohn. 2012. 'Two proposals'. *Cambridge Anthropology* 30 (2): 136–46.

Fullbrook, Edward. 2008. *Ontology and Economics: Tony Lawson and His Critics*. London & New York: Routledge.

Gad, Christopher and Casper Bruun Jensen. 2010. 'On the consequences of post-ANT'. *Science, Technology and Human Values* 19: 55–80.

Gad, Christopher, Casper Bruun Jensen and Brit Ross Winthereik. 2015. 'Practical ontology: Worlds in STS and anthropology'. *NatureCulture* 3:1–24.

Gatt, Caroline. 2013. 'Vectors, direction of attention and unprotected backs: re-specifying relations in anthropology'. *Anthropological Theory* 13 (4): 347–69.

Geertz, Clifford. 1973. *The Interpretation of Cultures*. New York: Basic Books.

Bibliography

1983. *Local Knowledge: Further Essays in Interpretive Anthropology.* New York: Basic Books.

Geismar, Haidy. 2011. '"Material Culture Studies" and other ways to theorize objects: a primer to a regional debate'. *Comparative Studies in Society and History* 53(1): 210–18.

Gell, Alfred. 1992. *The Anthropology of Time: Cultural Constructions of Temporal Maps and Images.* Oxford: Berg.

1998. *Art and Agency: An Anthropological Theory.* Oxford: Clarendon Press.

1999. 'Strathernograms, or the semiotics of mixed metaphors', in E. Hirsch (ed.), *The Art of Anthropology: Essays and Diagrams.* LSE Monographs on Social Anthropology, Volume 67, pp. 29–75. London: Bloomsbury Publishing.

Geschiere, Peter. 1997. *The Modernity of Witchcraft: Politics and the Occult in Postcolonial Africa.* Charlottesville: University Press of Virginia.

Gibson, James J. 1979. *The Ecological Approach to Visual Perception.* Boston: Houghton Mifflin.

Graeber, David. 2015. 'Radical alterity is just another way of saying "reality": a reply to Eduardo Viveiros de Castro'. *Hau: Journal of Ethnographic Theory* 5 (2): 1–41.

Green, Sarah F. 2005. *Notes from the Balkans: Locating Marginality and Ambiguity on the Greek-Albanian Border.* Princeton, NJ and Oxford: Princeton University Press.

2014. 'Anthropological knots: Conditions of possibilities and interventions'. *HAU: Journal of Ethnographic Theory* 4 (3): 1–21.

Gupta, Akhil and James Ferguson (eds.). 1997. *Culture, Power, Place: Explorations in Critical Anthropology.* Durham, NC: Duke University Press.

Hacking, Ian. 1999. *The Social Construction of What?* Cambridge, MA: Harvard University Press.

2002. *Historical Ontology.* Cambridge, MA: Harvard University Press.

Hage, Ghassan. 2012. 'Critical anthropological thought and the radical political imaginary today'. *Critique of Anthropology* 32 (3): 285–308.

2014. 'Critical anthropology as a permanent state of first contact'. *Cultural Anthropology* website, http://culanth.org/fieldsights/473-critical-anthropology-as-a-permanent-state-of-first-contact (accessed 20 June 2015).

Hallowell, Alfred I. 1960. 'Ojibwa ontology, behavior, and world view', in S. Diamond (ed.), *Culture in History. Essays in Honor of Paul Radin*, pp. 19–52. New York: Columbia University Press.

Hannerz, Ulf. 1993. *Cultural Complexity. Studies in the Social Organization of Meaning.* New York: Colombia University Press.

Harman, Graham. 2009. *Prince of Networks: Bruno Latour and Metaphysics.* Melbourne: re.press.

2010. *Towards Speculative Realism: Essays and Lectures*. Hants, UK: Zero Books.

2011. *The Quadruple Object*. Hants, UK: Zero Books.

Haraway, Donna J. 1989. *Primate Visions: Gender, Race, and Nature in the World of Modern Science*. New York: Routledge.

1991. 'A cyborg manifesto: science, technology and socialist feminism in the late twentieth century', in D. Haraway (ed.), *Simians, Cyborgs and Women: The Reinvention of Nature*, pp. 149–81. London: Free Association Books.

1992. 'The promises of monsters: a regenerative politics for inappropriate/d others', in L. Grossberg, C. Nelson and P. Treichler (eds.), *Cultural Studies*, pp. 295–337. New York: Routledge.

Harris, Oliver J. T. and John Robb. 2012. 'Multiple ontologies and the problem of the body in history'. *American Anthropologist* 114 (4): 668–79.

Hart, Keith. 2001. 'Cultural critique in Anthropology', in N. J. Smelser and P. B. Baltes (eds.), *International Encyclopedia of the Social and Behavioral Sciences* vol. 5, pp. 3037–41. New York: Elsevier.

Harvey, Graham. 2005. *Animism: Respecting the Living World*. London: Hurst & Co.

Hastrup, Frida. 2011. 'Shady plantations: theorizing coastal shelter in Tamil Nadu'. *Anthropological Theory* 11 (4): 435–9.

Hastrup, Kirsten. 1990. 'The ethnographic present: a reinvention'. *Cultural Anthropology* 5 (1): 45–61.

2004. 'Getting it right: knowledge and evidence in anthropology'. *Anthropological Theory* 4 (4): 455–72.

Hays, Terence E. 1986. 'Sacred flutes, fertility and growth in the papua New Guinea highlands'. *Anthropos* 81: 435–53.

Hegel, George W. F. 2006. *Hegel: Lectures on the Philosophy of Religion One-Volume Edition: The Lectures of 1827*. Oxford: Oxford University Press.

Heidegger, Martin. 1996. *Being and Time: A Translation of Zein und Zeit*. Translated by Joan Stambaugh. New York: State University of New York Press.

Helmreich, Steven. 2009. *Alien Ocean: Anthropological Voyages in Microbial Seas*. Berkeley, CA: University of California Press.

2012. 'Extraterrestrial relativism'. *Anthropological Quarterly* 85 (4): 1115–39.

Henare, Amiria. 2007. '*Taonga Māori*: encompassing rights and property in New Zealand', in A. Henare, M. Holbraad and S. Wastell (eds.), *Thinking Through Things: Theorizing Artefacts Ethnographically*, pp. 47–67. London: Routledge.

Henare, Amiria, Martin Holbraad and Sari Wastell (eds.). 2007. *Thinking Through Things: Theorising Artifacts Ethnographically*. London: Routledge.

Heywood, Paolo. 2012. 'Anthropology and what there is: reflections on "ontology"'. *Cambridge Anthropology* 30: 143–51.

2015. 'Equivocal locations: being "red" in "Red Bologna"'. *JRAI* 21: 855–71.

Bibliography

Hirsch, Eric. 2008. 'God or tidibe? Melanesian Christianity and the problem of wholes'. *Ethnos* 73: 141–62.

2014. 'Melanesian ethnography and the comparative project of anthropology: reflection on Strathern's analogical approach'. *Theory, Culture & Society* 31(2–3): 39–64.

Hirsch, Eric, and Marilyn Strathern (eds.). 2004. *Transactions and Creations: The Property Debates and the Stimulus of Melanesia*. Oxford: Berghahn Books.

Hobsbawm, Eric J. and Terence O. Ranger (eds.). 1983. *The Invention of Tradition*. Cambridge: Cambridge University Press.

Hodges, Matt. 2008. 'Rethinking time's arrow. Bergson, Deleuze and the anthropology of time'. *Anthropological Theory* 8 (4): 399–429.

Hogsden, Carl and Amiria Salmond. 2016. 'Ghosts in the Machine?'. Theorizing the Contemporary. *Cultural Anthropology* website. Available at https://culanth.org/fieldsights/830-ghosts-in-the-machine (accessed 01 September 2016).

Holbraad, Martin. 2003. 'Estimando a necessidade: os oráculos de ifá e a verdade em Havana'. *Mana* 9(2): 39–77.

2004. 'Response to Bruno Latour's "Thou shall not freeze-frame"'. Available at: https://sites.google.com/a/abaetenet.net/nansi/abaetextos/response-to-bruno-latours-thou-shall-not-freeze-frame-martin-holbraad (accessed 15 February 2016).

2005. 'Expending multiplicity: money in Cuban Ifá cults'. *Journal of the Royal Anthropological Institute (N.S)* 11 (2): 231–54.

2007. 'The power of powder: multiplicity and motion in the divinatory cosmology of Cuban Ifá (or Mana, Again)', in A. Henare, M. Holbraad, and S. Wastell (eds.), *Thinking Through Things: Theorising Artefacts Ethnographically*, pp. 189–225. London: Routledge.

2009. 'Ontology, ethnography, archaeology: an afterword on the ontography of things'. *Cambridge Archaeological Journal* 19 (3): 431–44.

2010. 'The whole beyond holism: gambling, divination and ethnography in Cuba,' in T. Otto and N. Bubandt (eds.), *Experiments in Holism: Theory and Practice in Contemporary Anthropology*, pp. 67–85. Oxford: Blackwell.

2012. *Truth in Motion: The Recursive Anthropology of Cuban Divination*. Chicago: University of Chicago Press.

2013a. 'Scoping recursivity: a comment on Franklin and Napier'. *Cambridge Anthropology* 31 (2): 123–7.

2013b. '*Revolución o muerte*: self-sacrifice and the ontology of Cuban revolution'. *Ethnos*. 79 (3): 365–87.

In press a. 'How myths make men in Afro-Cuban divination', in P. Pitarch and J.A. Kelly (eds.), *The Culture of Invention in the Americas*. Cambridge: Sean Kingston Publishing.

In press b. 'The contingency of concepts: transcendental deduction and ethnographic expression in anthropological thinking', in P. Charbonier, G. Salmon and P. Skafish (eds.), *Comparative Metaphysics: Ontology after Anthropology*. London: Rowman & Littlefield International.

Holbraad, Martin and Morten Axel Pedersen. 2009. 'Planet M: The intense abstraction of Marilyn Strathern'. *Anthropological Theory* 9 (4): 371–94.

2012. 'Revolutionary securitization: an anthropological extension of securitization theory'. *International Theory* 4 (2): 165–97.

(eds.). 2013. *Times of Security: Ethnographies of Fear, Protest and the Future*. London: Routledge.

Holbraad, Martin, Morten Axel Pedersen and Eduardo Viveiros de Castro. 2014. 'The politics of ontology: anthropological positions'. *Cultural Anthropology* website, http://culanth.org/fieldsights/462-the-politics-of-ontology-anthropological-positions (accessed 4 June 2015).

Holbraad, Martin and Rane Willerslev. 2007. 'Afterword: transcendental perspectivism: anonymous viewpoints from Inner Asia'. *Inner Asia* 9 (2): 329–45.

Honderich, Ted (ed.). 1995. *The Oxford Companion to Philosophy*. Oxford: Oxford University Press.

Howe, Cymene and Pandian, Anand. 2016. 'Lexicon for an anthropocene yet unseen'. Theorizing the Contemporary. *Cultural Anthropology* website, www.culanth .org/fieldsights/803-lexicon-for-an-anthropocene-yet-unseen (accessed 15 February 2016).

Howell, Signe. 1989. *Society and Cosmos: Chewong of Peninsular Malaysia*. Chicago: University of Chicago Press.

Højer, Lars. 2009. 'Absent powers: magic and loss in postsocialist Mongolia'. *Journal of the Royal Anthropological Institute (N.S)* 15 (3): 575–91.

Højer, Lars and Morten Axel Pedersen. In press. *Urban Hunters. Dealing and Dreaming in Times of Transition*. New Haven: Yale University Press.

Humphrey, Caroline. 1996. *Shamans and Elders. Experience, Knowledge, and Power among the Daur Mongols*. Oxford: Clarendon Press.

2007 'Inside and outside the mirror: Mongolian shaman's mirrors as instruments of perspectivism'. *Inner Asia* 9 (2): 173–96.

2008. 'Reassembling individual subjects. Events and decisions in troubled times'. *Anthropological Theory* 8 (4): 357–80.

Hutchins, Edwin. 1995. *Cognition in the Wild*. Cambridge, MA: The MIT Press.

Ingold, Tim. 1992. 'Editorial'. *Man* 27(1): 694–7.

1996. 'Hunting and gathering as ways of perceiving the environment'. In *Redefining Nature: Ecology, Culture, and Domestication* (eds.) R. Ellen & K. Fukui. Oxford: Berg.

Bibliography

1997. 'Eight themes in the anthropology of technology'. *Social Analysis* 41(1): 106–38.

1998. 'Totemism, animism, and the depiction of animals'. In *Animal. Anima. Animus* (eds.) M. Seppälä, J-P. Vanhala & L. Weintraub. Pori: Pori Art Museum.

2000. *The Perception of the Environment. Essays in Livelihood, Dwelling and Skill.* London: Routledge.

2007. 'Materials against materiality'. *Archaeological Dialogues* 14 (1): 1–16.

2008. 'When ANT meets SPIDER: Social theory for arthropods', in C. Knappet and L. Malafouris (eds.), *Material Agency: Towards a Non-Anthropocentric Approach*, pp. 209–15. New York: Springer.

2014. 'That's enough about ethnography!'. *Hau: Journal of Ethnographic Theory* 4 (1): 383–95.

Ishii, Miho. 2012. 'Acting with things: self-poiesis, actuality, and contingency in the formation of divine worlds'. *HAU: Journal of Ethnographic Theory* 2 (2): 371–88.

Jackson, Michael (ed.). 1996. *Things as They Are. New Directions in Phenomenological Anthropology.* Bloomington: Indiana University Press.

1998. *Minima Ethnographica: Intersubjectivity and the Anthropological Project.* Chicago: University of Chicago Press.

Jensen, Casper Bruun 2004. 'A nonhumanist disposition: on performativity, practical ontology, and interventions'. *Configurations* 12 (2): 229–61.

2012. 'Proposing the motion: "the task of anthropology is to invent relations"'. *Critique of Anthropology* 32: 47–53.

2013. 'Two forms of the outside. Castaneda, Blanchot, ontology'. *HAU: Journal of Ethnographic Theory* 3 (3): 309–35.

2014a. 'Practical Ontologies.' Theorizing the Contemporary, *Cultural Anthropology* website. https://culanth.org/fieldsights/466-practical-ontologies (accessed 13 January 2014).

2014b. 'Experiments in good faith and hopelessness. Toward a postcritical social science'. *Common Knowledge* 20 (2): 337–362.

2015. 'Experimenting with political materials: environmental infrastructure and ontological transformations'. *Distinktion: Scandinavian Journal of Social Theory* 16 (1): 17–30.

In press. 'Multinatural infrastructures: the many natures and cultures of Phnom Penh's sewage system', in P. Harvey, C. B. Jensen and A. Morita (eds.), *Infrastructures and Social Complexity: A Routledge Companion.* London and New York. Routledge.

Jensen, Casper Bruun and Kjetil Rödje (eds.). 2009. *Deleuzian Intersections in Science, Technology and Anthropology.* Oxford: Berghahn.

Jensen, Casper Bruun, Morten Axel Pedersen and Brit Ross Winthereik (eds.). 2011. 'Comparative Relativism', special issue of *Common Knowledge* 17 (1).

Jensen, Casper Bruun and Atsuro Morita (eds.). 2012. 'Anthropology as critique of reality: a Japanese turn'. Forum in *HAU: Journal of Ethnographic Theory* 2 (2): 358–405.

Jensen, Casper Bruun and Brit Ross Winthereik. 2013. *Monitoring Movements in Development Aid: Recursive Infrastructures and Partnerships*. Cambridge, MA: MIT Press.

Jolly Margaret. 1992. 'Partible persons and multiple authors'. *Pacific Studies* 15 (1): 137–48.

Josephides, Lisette. 1991. 'Metaphors, metathemes, and the construction of sociality: a critique of the new Melanesian ethnography'. *Man (N.S.)* 26: 145–61.

Kapferer, Bruce. 1991. *A Celebration of Demons. Exorcism and the Aesthetics of Healing in Sri Lanka*. Oxford: Berg.

2010. *Legends of People, Myths of State: Violence, Intolerance, and Political Culture in Sri Lanka and Australia*. Oxford: Berghahn Press.

Keane, Webb. 2007. *Christian Moderns. Freedom and Fetish in the Mission Encounter*. Berkeley, CA: University of California Press.

2013. 'Ontologies, anthropologists, and ethical lives'. *Hau: Journal of Ethnographic Theory* 3 (1): 186–91.

Keesing, Roger. 1992. Review of 'Marilyn Strathern "The Gender of the Gift: Problems with Women and Problems with Society in Melanesia"'. *Pacific Studies* 15 (1): 129–37.

Kelly, Anne. 2011. 'Entomological extensions: model huts and fieldworks', in J. Edwards and M. Petrovic-Steger (eds.), *Recasting Anthropological Knowledge: Inspiration and Social Science*, pp.70–87. Cambridge: Cambridge University Press.

Kelly, John (ed.). 2014. *The Ontological French Turn*. Colloquium in *HAU: Journal of Ethnographic Theory* 4 (1): 259–360.

Kelly, José Antonio. 2005. 'Fractality and the exchange of perspectives', in M. S. Mosko, and H. D. Frederick (eds.), *On the Order of Chaos. Social Anthropology and the Science of Chaos*, pp. 108–35. Oxford: Berghahn Books.

2011. *State Healthcare and Yanomami Transformations: A Symmetrical Ethnography*. Tucson, AZ: University of Arizona Press.

Kerridge, Ian. 2003. 'Altruism or reckless curiosity? A brief history of self-experimentation in medicine'. *Internal Medicine Journal* 33: 203–7.

Knappett, Carl and Lambros Malafouris (eds.). 2008. *Material Agency: Towards a Non-anthropocentric Approach*. New York: Springer.

Knox, Hannah and Antonia Walford (eds.). 2016. 'Is there an ontology to the digital?' *Cultural Anthropology* website, http://culanth.org/conversations/17-theorizing-the-contemporary (accessed 1 September 2016).

Knudsen, Are J. 1998. 'Beyond cultural relativism? Tim Ingold's "ontology of dwelling"'. *CMI Working Paper WP* 1998: 7. Bergen: Chr. Michelsen Institute.

Bibliography

Kohn, Eduardo. 2008. 'How dogs dream: Amazonian natures and the politics of trans-species engagement'. *American Ethnologist* 34 (1): 3–24.

— 2013. *How Forests Think: Toward an Anthropology Beyond the Human*. London & Berkeley, CA: University of California Press.

— 2014. 'What an ontological anthropology might mean'. *Cultural Anthropology* website, http://culanth.org/fieldsights/463-what-an-ontological-anthropology-might-mean (accessed 15 June 2015).

— 2015. 'Anthropology of ontologies'. *Annual Review of Anthropology* 44: 311–27.

Krauss, Rosalind. 1986. *The Originality of the Avant-Garde and Other Modernist Myths*. Cambridge, MA: MIT Press.

Kristensen, Benedikte. 2007. 'The human perspective'. *Inner Asia* 9 (2): 275–90.

Krøijer, Stine. 2015. *Figurations of the Future: Forms and Temporality of Left Radical Politics in Northern Europe*. Oxford: Berghahn Books.

Kuper, Adam. 2000. *Culture: The Anthropologists' Account*. Cambridge, MA: Harvard University Press.

Kuper, Hilda. 1984. 'Function, history, biography: reflections on fifty years in the British anthropological tradition', in G. Stocking (ed.), *Functionalism Historicized. Essays on British Social Anthropology*, pp. 192–213. Madison, WI: University of Wisconsin Press.

Küchler, Susanne. 1988. 'Malangan: objects, sacrifice, and the production of memory'. *American Ethnologist* 15 (4): 625–37.

— 2002. *Malanggan: Art, Memory and Sacrifice*. New York: Berg.

Laidlaw, James. 2012. 'Ontologically challenged'. *Anthropology of this Century* 4, Available at: http://aotcpress.com/articles/ontologically-challenged (accessed 22 April 2015).

— 2013. *The Subject of Virtue: An Anthropology of Ethics and Freedom*. Cambridge: Cambridge University Press.

Laidlaw, James and Paolo Heywood. 2013. 'One more turn and you're there'. *Anthropology of This Century*, vol. 7, London, May 2013. Available at: http://aotcpress.com/articles/turn/ (accessed 15 January 2016).

Latour, Bruno. 1988. *Irréductions*. Published with *The Pasteurization of France*. Cambridge, MA: Harvard University Press.

— 1990. 'The force and the reason of experiment', in H. E. Le Grand (ed.), *Experimental Inquiries*, pp. 49–80. Dordrecht, Netherlands: Kluwer Academic Publishers.

— 1993. *We Have Never Been Modern*. Translated by Catherine Porter. London: Prentice-Hall.

— 1999. *Pandora's Hope: Essays on the Reality of Science Studies*. Cambridge, MA: Harvard University Press.

2004a. 'Whose cosmos, whose cosmopolitics? Comments on the peace terms of Ulrich Beck'. *Common Knowledge* 10 (3): 450–92.

2004b. '"Não congelarás a imagem", ou: como não desentender o debate ciência-religião'. *Mana* 10 (2): 349–76.

2004c. Why has critique run out of steam? From matters of fact to matters of concern. *Critical Inquiry* 30 (2): 225–48.

2005. *Reassembling the Social: An Introduction to Actor-Network-Theory.* Oxford: Oxford University Press.

2009. 'Perspectivism: type or bomb?' *Anthropology Today* 25 (2): 1–2.

2010a. *On the Modern Cult of the Factish Gods.* Durham, NC and London: Duke University Press.

2010b. 'An attempt at a "compositionist manifesto"'. *New Literary History* 41: 471–90.

2013. *An Inquiry into Modes of Existence: An Anthropology of the Moderns.* Translated by Cathy Porter. Cambridge, MA.: Harvard University Press.

2014a. 'What is the style of matters of concern', in N. Gaskill and A. J. Nocek (eds.), *The Lure of Whitehead*, pp. 92–127. Minnesota: University of Minnesota Press.

2014b. 'Agency at the time of the anthropocene'. *New Literary History* 45 (1): 1–18.

Latour, Bruno, and Steve Woolgar. 1986. *Laboratory Life: The Construction of Scientific Facts.* Princeton, NJ: Princeton University Press.

Latour, Bruno, Pablo Jensen, Tommaso Venturini, Sébastian Grauwin and Dominique Boullier. 2012. 'The whole is always smaller than the parts – a digital test of Gabriel Tarde's monads'. *British Journal of Sociology* 63 (4): 590–615.

Law, John. 2000. 'Transitivities'. *Society and Space* 18: 133–48.

2004. *After Method: Mess in Social Science Research.* London and New York: Routledge.

Law, John, and Annemarie Mol (eds.). 2002. *Complexities: Social Studies of Knowledge Practices.* Durham: Duke University Press.

Lawson, Tony. 2012. 'Ontology and the study of social reality: emergence, organisation, community, power, social relations, corporations, artefacts and money'. *Cambridge Journal of Economics* 36 (2): 345–85.

Layton, Robert. 2003. 'Art and Agency: a reassessment'. *Journal of the Royal Anthropological Institute* 9 (3): 447–64.

Leach, James. 2003. *Creative Land: Place and Procreation on the Rai Coast of Papua New Guinea.* New York: Berghahn Books.

2004. 'Modes of creativity', in E. Hirsch and M. Strathern (eds.), *Transactions and Creations: The Property Debates and the Stimulus of Melanesia*, pp.151–75. Oxford: Berghahn Books.

Bibliography

2007. 'Differentiation and encompassment: a critique of Alfred Gell's theory of the abduction of creativity', in Henare et al. (eds.), *Thinking Through Things: Theorising artefacts ethnographically*, pp. 167–88. London: Routledge.

2014. 'Choreographic Objects: Contemporary dance, digital creations and prototyping social visibility'. *Journal of Cultural Economy* 7 (4): 458–75.

Lebner, Ashley. 2016. La redescription de l'anthropologie selon Marilyn Strathern. *L'Homme* 218.

Lemonnier, Pierre. 1992. *Elements for an Anthropology of Technology*. Ann Arbor, MI: University of Michigan, Museum of Anthropology.

Lenclud, Gérard, Stefan Helmreich, Stephan Feuchtwang, Bruce Kapferer, Christina Toren, Michael Lambek, Marcela Coelho de Souza and Philippe Descola. 2014. 'Book Symposium – Beyond nature and culture (Philippe Descola)'. *Hau: Journal of Ethnographic Theory* 4 (3): 363–443.

Lévi-Strauss, Claude. 1954. 'The mathematics of man'. *International Social Science Bulletin* 6: 581–90.

1963. *Structural Anthropology*. Translated by Claire Jacobson and Brooke Grundfest Schoepf. New York: Basic Books.

1964. *Totemism*. Translated by R. Needham. London: Merlin Press.

1966. *The Savage Mind*. Translated by George Weidenfield and Nicholson Ltd. Chicago: Chicago University Press.

1969. *The Raw and the Cooked: Introduction to a Science of Mythology*. Translated by John and Doreen Weightman. New York: Harper & Row.

1979. *From Honey to Ashes*. Translated by John Weightman and Doreen Weightman. London: Octagon Press.

1990a. *The Origin of Table Manners*. Chicago: University of Chicago Press.

1990b. *The Naked Man*. Chicago: University of Chicago Press.

Lien, Marianne and John Law. 2011. '"Emergent Aliens": on salmon, nature, and their enactment'. *Ethnos* 76 (1): 65–87.

2012. 'Slippery: field notes on empirical ontology'. *Social Studies of Science*. 43 (3): 363–78.

Lilla, Mark. 2008. *The Stillborn God: Religion, Politics, and the Modern West*. New York: Vintage Books.

Lima, Tania Stolze. 1999. 'The two and its many: reflections on perspectivism in a Tupi cosmology'. *Ethnos* 64 (1): 107–31.

Lloyd, Geoffrey E. R. 2007. *Cognitive Variations*. Oxford: Clarendon Press.

2011. 'Humanity between gods and beasts? Ontologies in question'. *Journal of the Royal Anthropological Institute (N.S.)* 17: 829–45.

2012. *Being, Humanity, and Understanding*. Oxford: Oxford University Press.

Luhrmann, Tanya M. 2012. *When God Talks Back: Understanding the American Evangelical Relationship with God*. New York: Alfred A. Knopf.

2013. 'What anthropology should learn from G.E.R. Lloyd'. *Hau: Journal of Ethnographic Theory* 3 (1): 171-3.

Luhrmann, Tanya M., Aparecida Vilaça, Steven Sangren, Webb Keane, Carlo Severi, James Laidlaw, Anne-Christine Taylor, and Geoffrey E. R. Lloyd. 2013. Book Symposium – Being, humanity, and understanding (G. E. R. Lloyd). *Hau: Journal of Ethnographic Theory* 3 (1): 171-209.

Lynch, Michael. 2013. 'Ontography: investigating the production of things, deflating ontology'. *Social Studies of Science* 43 (3): 444-62.

Lyons, Kristina Marie 2014. 'Soil science, development, and the "elusive nature" of Colombia's Amazonian Plains'. *The Journal of Latin American and Caribbean Anthropology* 19 (2): 212-36.

Malafouris, Lambros. 2013. *How Things Shape the Mind: A Theory of Material Engagement.* Cambridge, MA: MIT Press.

Malinowski, Bronislaw. 1961. *Argonauts of the Western Pacific.* New York: E. P. Dutton.

2002. *Argonauts of the Western Pacific: An Account of Native Enterprise and Adventure in the Archipelagoes of Melanesian New Guinea.* London: Routledge.

Mamet, David. 1997. *True and False: Heresy and Common Sense for the Actor.* New York: Pantheon Books.

Maniglier, Patrice. 2014. 'A metaphysical turn? Bruno Latour's "An Inquiry into Modes of Existence"'. *Radical Philosophy* 187: 37-44.

Marcoulatos, Iordanis. 2003. 'The secret life of things: rethinking social ontology'. *Journal for the Theory of Social Behaviour* 33 (3): 245-78.

Marcus, George. 1995. 'Ethnography in/of the world system: The emergence of multi-sited ethnography'. *Annual Review of Anthropology* 24: 95-117.

Marcus, George and Michael M. J. Fischer (eds.). 1986. *Anthropology as Cultural Critique: An Experimental Moment in the Social Sciences.* Chicago: University of Chicago Press.

Marcus, George and Michael M. J. Fischer. 1993. 'Ethnography in/of the world system: the emergence of multi-sited ethnography'. *Annual Review of Anthropology* 24: 95-117.

1986. *Anthropology as Cultural Critique: An Experimental Moment in the Human Sciences.* Chicago: University of Chicago Press.

Martin, Keir. 2013. *The Death of the Big Men and the Rise of the Big Shots: Custom and Conflict in Eastern New Britain.* Oxford: Berghahn Books.

Maurer, Bill. 2005. *Mutual Life, Limited. Islamic Banking, Alternative Currencies, Lateral Reason.* Princeton: Princeton University Press.

Mauss, Marcel. 1990. *The Gift: Forms and Functions of Exchange in Archaic Societies.* Translated by W. D. Halls. London: Routledge.

Mbembe, Achille. 1992. 'Provisional notes on the postcolony'. *Africa* 62 (1): 3-37.

Mead, Margaret. 1961. *Coming of Age in Samoa.* New York: Morrow Quill Paperbacks.

Bibliography

Meillassoux, Quentin. 2008. *After Finitude: An Essay on the Necessity of Contingency*. New York: Continuum.

Menary, Richard (ed.). 2010. *The Extended Mind*. Cambridge, MA: MIT Press/ Bradford

Meyer, Birgit. 1999. *Translating the Devil. Religion and Modernity among the Ewe in Ghana*. Edinburgh: Edinburgh University Press.

Miller, Daniel. 1987. *Material Culture and Mass Consumption*. Oxford: Basil Blackwell.

    2005. 'Materiality: an introduction', in D. Miller (ed.) *Materiality*, pp. 1–50. Durham, NC and London: Duke University Press.

    2007. 'Stone age or plastic age?' *Archaeological Dialogues* 14 (1): 23–7.

Mithen, Steven. 1996. *The Prehistory of the Mind. A Search for the Origins of Art, Religion and Science*. London: Thames and Hudson.

Miyazaki, Hirokazu. 2006. 'Economy of dreams: hope in global capitalism and its critiques'. *Cultural Anthropology* 21 (2): 147–72.

    2013. *Arbitraging Japan: Dreaming of Capitalism as the End of Finance*. Berkeley, CA: University of California Press.

Miyazaki, Hirokazu and Annelise Riles. 2005. 'Failure as an endpoint', in A. Ong and S. J. Collier (eds.), *Global Assemblages. Technology, Politics, and Ethics as Anthropological Problems*, pp. 320–32. Oxford: Blackwell.

Mol, Annemarie. 1999. 'Ontological politics: a word and some questions', in J. Law and J. Hassard (eds.), *Actor Network Theory and After*, pp. 74–89. Oxford and Keele: Blackwell and the Sociological Review.

    2002. *The Body Multiple: Ontology in Medical Practice*. Durham, NC and London: Duke University Press.

Moore, Henrietta. 1988. *Feminism and Anthropology*. Cambridge, UK: Polity Press.

Morita, Atsuro. 2013. 'The ethnographic machine: experimenting with context and comparison in Strathernian ethnography'. *Science, Technology, & Human Values* 39 (2): 214–35.

Morris, Rosalind C. 2007. 'Legacies of Derrida: anthropology'. *Review of Anthropology* 36: 355–89.

Morton, Timothy. 2013. *Hyperobjects Philosophy and Ecology after the End of the World*. Minneapolis, MN: University of Minnesota Press.

Mosko, Mark. 1985. *Quadripartite Structures. Categories, Relations, and Homologies in Bush Mekeo culture*. Cambridge: Cambridge University Press.

    1992. 'Motherless sons: "divine heroes" and "partible persons" in Melanesia and Polynesia'. *Man (N.S.)* 27: 697–717.

    2010. 'Partible penitents: dividual personhood and Christian practice in Melanesia and the West'. *Journal of the Royal Anthropological Institute* 16 (1): 215–40.

    2015. 'Unbecoming individuals. The partible character of the Christian person'. *Hau: Journal of Ethnographic Theory* 5 (1): 361–93.

Mosko, Mark S. and Frederick H. Damon. 2005. *On the Order of Chaos. Social Anthropology and the Science of Chaos*. Oxford: Berghahn Books.

Moutu, Andrew. 2013. *Names are Thicker than Blood: Kinship and Ownership amongst the Iatmul*. Oxford: Oxford University Press.

Munn, Nancy D. 1986. *The Fame of Gawa: A Symbolic Study of Value Transformation in a Massim (Papua New Guinea) Society*. Cambridge: Cambridge University Press.

Myhre, Knut C. 2013. 'Introduction: cutting and connecting – "Afrinesian" perspectives on networks, relationality, and exchange'. *Social Analysis* 57 (3): 1–24.

Nielsen, Morten. 2011. 'Futures within: reversible time and house-building in Maputo, Mozambique'. *Anthropological Theory* 11 (4): 397–423.

——— 2012. 'Interior swelling. On the expansive effects of ancestral interventions in Maputo, Mozambique'. *Common Knowledge* 18(3):433–50.

——— 2013. 'Analogic Asphalt: suspended value conversations among young road workers in Southern Mozambique'. *HAU: Journal of Ethnographic Theory* 3 (2): 79–96.

——— 2014. 'A wedge of time: futures in the present and presents without futures in Maputo, Mozambique'. *Journal of the Royal Anthropological Institute* 20 (S1): 166–82.

Nielsen, Morten and Morten Axel Pedersen. 2012. Concept paper 'An Anthropological Theory of Distortion'. Available at: http://anthropology .ku.dk/research/research-projects/current-projects/distortion/documents/ Distortion_Concept_Paper._January_2012.pdf (accessed 15 February 2016).

——— 2015. *Comedy of Things: 'About' and 'Backgroumd'*. Available at: http:// comedyofthings.com/background/ (Accessed 15 February 2016).

——— (eds.). Forthcoming. *The Comedy of Things: An Anthropolgical Experiment*. Brooklyn, NY: Punctum Books.

Nietzsche, Friedrich. 2001. *The Gay Science: With a Prelude in German Rhymes and an Appendix of Songs*. Edited by Bernard Williams, translated by Josefine Nauckhoff and Adrian Del Caro. Cambridge: Cambridge University Press.

Olsen, Bjørnar. 2003. 'Material culture after text: remembering things'. *Norwegian Archaeological Review* 36 (3): 87–104.

Olwig, Karen F. and Kirsten Hastrup (eds.). 1997. *Siting Culture. The Shifting Anthropological Object*. London: Routledge.

Ortner, Sherry.B. 1984. 'Theory in anthropology since the sixties'. *Comparative Studies in Society and History* 26 (1): 126–66.

Paleček, Martin and Mark Risjord. 2013. 'Relativism and the ontological turn within anthropology'. *Philosophy of the Social Sciences* 43 (1): 3–23.

Pandian, Anand. 2010. 'Interior horizons: an ethical space of selfhood in South India'. *Journal of the Royal Anthropological Institute (N.S.)* 16: 64–83.

Bibliography

Pedersen, Morten Axel. 2001. 'Totemism, animism and North Asian indigenous ontologies'. *Journal of the Royal Anthropological Institute* 7 (3): 411–27.

——— 2003. 'Networking the landscape. Place, power and decision making in Northern Mongolia', in A. Roebstorff, N. Bubandt and K. Kull (eds.), *Imagining Nature. Practises of Cosmology and Identity*, pp. 238–59. Aarhus: Aarhus University Press.

——— 2007. 'Talismans of thought. Shamanic ontology and extended cognition in Northern Mongolia', in A. Henare, M. Holbraad and S. Wastell (eds.), *Thinking Through Things. Theorizing Artefacts Ethnographically*, pp. 141–66. London: Routledge.

——— 2011. *Not Quite Shamans. Spirit Worlds and Political Lives in Northern Mongolia.* Ithaca, NY: Cornell University Press.

——— 2012a. 'Common nonsense: a review of certain recent reviews of "the ontological turn"'. *Anthropology of This Century*, 5. Available at: http://aotcpress.com/articles/common_nonsense/ (accessed 15 February 2016).

——— 2012b. 'The task of anthropology is to invent relations: for the motion'. *Critique of Anthropology* 32 (1): 59–65

——— 2012c. 'A day in the Cadillac: the work of hope in urban Mongolia'. *Social Analysis* 56 (2): 136–51.

——— 2013a. 'Islands of nature: insular objects and frozen sprits in Northern Mongolia', in K. Hastrup (ed.), *Anthropology and Nature*, pp. 96–107. London: Routledge.

——— 2013b. 'The fetish of connectivity', in P. Harvey, E. C. Casella, G. Evans, H. Knox, C. Mclean, E. B. Silva, N. Thoburn and K. Woodward (eds.), *Objects and Materials. A Routledge Companion*, pp. 197–207. London: Routledge.

——— 2013c. 'Taking things too seriously'. *Professorial inaugural lecture*, Department of Anthropology, University of Copenhagen, 6 September 2013.

——— 2014. 'Shamanic spirits in transition: postsocialism as political cosmology', in A. Abrahamson and M. Holbraad (eds.), *Framing Cosmologies: The Anthropology of Worlds*, pp. 161–84. Manchester: Manchester University Press.

——— 2017. "The politics of paradox. Kierkegaardian theology and national conservatism in Denmark" In N. Rapport (ed.), *Distortion: Social Processes beyond the Structured and the Systemic*. London: Routledge.

Pedersen, Morten Axel, Rebecca Empson and Caroline Humphrey (eds.). 2007. 'Inner Asian perspectivism'. Special issue of *Inner Asia* 10 (2).

Pedersen, Morten Axel and Mikkel Bunkenborg. 2012. 'Roads that separate: Sino-Mongolian relations in the Inner Asian desert'. *Mobilities* 7 (4): 554–69.

Pedersen, Morten Axel and Lars Højer. 2008. 'Lost in transition: fuzzy property and leaky selves in Ulaanbaatar'. *Ethnos* 73: 73–96.

Pedersen, Morten Axel and Morten Nielsen. 2013. 'Trans-temporal hinges reflections on a comparative ethnographic study of Chinese infrastructural projects in Mozambique and Mongolia'. *Social Analysis* 57 (1): 122–42.

Pedersen, Morten Axel and Rane Willerslev. 2012. '"The soul of the soul is the body": rethinking the soul through North Asian ethnography'. *Common Knowledge* 18 (3): 464–86.

Peirce, Charles Sanders. 1931–35. *Collected Papers of Charles Sanders Peirce*. Cambridge, MA: Harvard University Press.

Pickering, Andrew. 1995. *The Mangle of Practice: Time, Agency and Science*. Chicago and London: University of Chicago Press.

      2011. 'Ontological politics: realism and agency in science, technology and art'. *Insights* 4 (9): 2–11.

      2016. 'The ontological turn: taking different worlds seriously'. Available at: https://architecture.mit.edu/sites/architecture.mit.edu/files/attachments/lecture/tokyo-rev-060815.pdf (accessed 31 August 2016)

Pickering, Andrew and Keith Guzik. 2008. *The Mangle in Practice: Science, Society, and Becoming*. Durham, NC: Duke University Press.

Pickles, Anthony. 2013. 'Pocket calculator: a humdrum "obviator" in Papua New Guinea?'. *Journal of the Royal Anthropological Institute* 19 (3): 510–26.

Piette, Albert. 2012. *De l'ontologie en anthropologie*, Paris: Berg International.

      2015. 'God and the anthropologist: the ontological turn and human-oriented anthropology'. *Tsantsa* 20: 97–107.

Pina-Cabral, João de. 2014. 'World: an anthropological examination (part 1)'. *HAU: Journal of Anthropological Theory* 4 (1): 49–73.

Pinney, Chris. 2005. 'Things happen: or, from which moment does that object come?', in D. Miller (ed.), *Materiality*, pp. 256–72. Durham, NC and London: Duke University Press.

Pinney, Christopher and Nicholas Thomas (eds.). 2001. *Beyond Aesthetics: Art and the Technologies of Enchantment*. Oxford: Berg.

Pitarch, Pedro. 2011. 'The two Maya bodies: an elementary model of Tzeltal personhood'. *Ethnos* 77 (1): 93–114.

Povinelli, Elizabeth. 2002. *The Cunning of Recognition: Indigenous Alterities and the Making of Australian multiculturalism*. London and Durham, NC: Duke University Press.

      2012. 'The will to be otherwise / the effort of endurance'. *South Atlantic Quarterly* 111 (3): 453–7.

      2014. 'Geontologies of the Otherwise'. Theorizing the Contemporary. *Cultural Anthropology* website, www.culanth.org/fieldsights/465-geontologies-of-the-otherwise (accessed 15 February 2016).

Puett, Michael. 2014. 'Ritual disjunctions: ghosts, anthropology, and philosophy', in V. Das, M. Jackson, A. Kleinman and B. Singh (eds.), *The Ground Between: Anthropologists Engage Philosophy*, pp. 218–33. Durham, NC: Duke University Press.

Bibliography

Putnam, Hilary. 1975. 'The meaning of "meaning".' *Mind, Language and Reality: Philosophical Papers 7*: 131–93. Cambridge: Cambridge University Press.

Radcliffe-Brown, Alfred R. 1958. *Method in Social Anthropology.* Edited by M. N. Srinivas. Chicago: Chicago University Press.

Rabinow, Paul. 1977. *Reflections on Fieldwork in Morocco.* Berkeley, CA: University of California Press.

Ramos, Alcida R. 2012. 'The politics of perspectivism.' *Annual Review of Anthropology* 41: 481–94.

Rappaport, Roy A. 1999. *Ritual and Religion in the Making of Humanity.* Cambridge: Cambridge University Press.

Rapport, Nigel. 2007. 'An outline for cosmopolitan study: reclaiming the human through introspection.' *Current Anthropology* 48 (2) 257–83.

Ratner, Helene. 2012. *Promises of Reflexivity: Managing and Researching Inclusive Schools.* PhD thesis. Copenhagen Business School.

Reed, Adam. 2004. *Papua New Guinea's Last Place. Experiences of Constraint in a Postcolonial Prison.* Oxford: Berghahn Books.

——— 2007. 'Smuk is king: the action of cigarettes in a Papua New Guinea prison', in A. Henare, M. Holbraad and S. Wastell (eds.), *Thinking Through Things. Theorizing Artefacts Ethnographically*, pp. 32–46. London: Routledge.

——— 2011. 'Inspiring Strathern', in J. Edwards and M. Petrovic-Steger (eds.), *Recasting Anthropological Knowledge: Inspiration and Social Science*, pp. 165–82. Cambridge: Cambridge University Press.

Rheinberger, Hans. 1994. 'Experimental systems: historiality, narration, and deconstruction.' *Science in Context* 7 (1): 65–81.

Ricart, Ender. 2014. 'Field of difference: limitations of the political in ontological anthropology.' Fieldsights – Commentary. Cultural Anthropology website, http://culanth.org/fieldsights/524-field-of-difference-limitations-of-the-political-in-ontological-anthropology (accessed 22 April 2015).

Ricoeur, Paul. 1998. 'Thinking creation', in A. LaCocque and P. Ricoeur, *Thinking Biblically: Exegetical and Hermeneutical Studies*, pp. 31–67. Translated by David Pellauer. Chicago: University of Chicago Press.

Riles, Annelise. 1998. 'Infinity within the brackets.' *American Ethnologist* 25 (3): 378–398.

——— 2001. *The Network Inside Out.* Michigan: University of Michigan Press.

Rio, Knut Mikjel. 2007. *The Power of Perspective. Social Ontology and Agency on Ambrym Island, Vanuatu.* Oxford: Berghahn Books.

Robbins, Joel. 2004a. *Becoming Sinners. Christianity and Moral Torment in a Papua New Guinea Society.* Berkeley, CA: University of California Press.

——— 2004b. 'The globalization of pentecostal and charismatic Christianity.' *Annual Review of Anthropology* 33: 117–43.

2007. 'Continuity thinking and the problem of Christian culture. belief, time, and the Anthropology of Christianity'. *Current Anthropology* 48 (1): 5–17.

2010. 'Melanesia, Christianity, and cultural change: a comment on Mosko's "Partible penitents"'. *Journal of the Royal Anthropological Institute* 16 (2): 241–3.

2012. 'Transcendence and the anthropology of Christianity: language, change and individualism'. *Suomen Antropologi: The Journal of the Finnish Anthropological Society* 37 (2): 5–23.

2015. 'Dumont's hierarchical dynamism. Christianity and individualism revisited'. *Hau: Journal of Ethnographic Theory* 5 (1): 173–95.

Robbins, Joel, Bambi B. Schieffelin and Aparecida Vilaça. 2014. 'Evangelical conversion and the transformation of the self in Amazonia and Melanesia: Christianity and the revival of anthropological comparison'. *Comparative Studies in Society and History* 56 (3): 559–90.

Ruel, Malcolm. 1982. 'Christians as believers', in J. Davis (ed.), *Religious Organization and Religious Experience*, pp. 9–31. London: Academic Press.

Sahlins, Marshall. 1974. *Stone Age Economics*. Chicago: Aldine Transaction.

1985. *Islands of History*. Chicago: University of Chicago Press.

1999. 'Two or three things that I know about culture'. *Journal of the Royal Anthropological Institute (N.S.)* 5 (3): 399–421.

Said, Edward. 1978. *Orientalism*. New York: Vintage Books.

Salmond, Amiria. 2013. 'Transforming translations (part 1): the owner of these bones'. *HAU: Journal of Ethnographic Theory* 3 (3): 1–32.

2014. 'Transforming translations (part 2): addressing ontological alterity'. *HAU: Journal of Ethnographic Theory* 4 (1): 155–87.

Forthcoming. Uncommon things. In: Marisol de la Cadena and Mario Blaser (eds.) *Dialogues for the Reconstitution of Worlds*. Durham and London: Duke University Press.

Sangren, Steven. 1988. 'Rhetoric and the authority of ethnography: "postmodernism" and the social reproduction of texts'. *Current Anthropology* 29 (3): 405–35.

2007. 'Anthropology of anthropology? Further reflections on reflexivity', *Anthropology Today* 23(4): 13–16.

Schlecker, Markus and Eric Hirsch. 2001. 'Incomplete knowledge: ethnography and the crisis of context in studies of media science and technology'. *History of the Human Sciences* 14 (1): 69–87.

Schmidt, Jessica. 2015. 'Worlds beyond critique: choking on the joke?'. *Global Dialogues* 10: 12–17.

Schmidt, Mario. 2015. 'Critique as a paratopical joke: the political apoliticalness of anthropology'. *Global Dialogues* 10: 18–24.

Schneider, David M. 1968. *American Kinship: A Cultural Account*. Chicago: University of Chicago Press.

Bibliography

1995. *Schneider on Schneider: The Conversion of the Jews and Other Anthropological Stories.* Edited by Richard Handler. Durham, NC: Duke University Press.

Schrempp, Gregory. 1992. *Magical Arrows: The Maori, Greeks, and the Folklore of the Universe.* Madison, WI: *University of Wisconsin Press.*

2012. *The Ancient Mythology of Modern Science: A Mythologist Looks (Seriously) at Popular Science Writing.* Montreal and Kingston: McGill-Queen's University Press.

Scott, Michael W. 2005. 'Hybridity, vacuity, and blockage: visions of chaos from anthropological theory, island Melanesia, and Central Africa'. *Comparative Studies in Society and History* 47 (1): 190–216.

2007. *The Severed Snake: Matrilineages, Making Place, and a Melanesian Christianity in Southeast Solomon Islands.* Durham, NC: Carolina Academic Press.

2013a. 'The anthropology of ontology (religious science?)'. *Journal of the Anthropological Institute (N.S).* 19 (4): 859–72.

2013b. 'Steps to a methodological non-dualism'. *Critique of Anthropology* 33 (3): 303–9.

2014. 'To be a wonder: anthropology, cosmology, and alterity', in A. Abramson and M. Holbraad (eds.), *The Cosmological Frame: The Anthropology of Worlds,* pp. 31–54. Manchester: Manchester University Press.

2015a. '"When people have a vision they are very disobedient": A Solomon Islands case study for the Anthropology of Christian ontologies', in M. Fuchs, A. Linkenbach-Fuchs, and W. Reinhard (eds.), *Individualisierung durch christliche Mission?,* pp. 635–50. Wiesbaden: Harrassowitz Verlag.

2015b. 'Cosmogony today: counter-cosmogony, perspectivism, and the return of anti-biblical polemic'. *Religion and Society: Advances in Research* 6 (1): 44–61.

Searle, John. 1995. *The Construction of Social Reality.* New York: Free Press.

2006. 'Social ontology: some basic principles'. *Anthropological Theory* 6 (1): 12–29.

Segal, Robert A. 1999. 'Durkheim in Britain: the work of Radcliffe-Brown'. *JASO* 30 (2): 131–62.

Shaffer, Simon. 1994. *From Physics to Anthropology – & Back Again.* Prickly Pears Pamphlets Series No. 3. Cambridge: Prickly Pears Pamphlets.

Shapin, Steven and Simon Schaffer. 1985. *Leviathan and the Air Pump: Hobbes, Boyle and the Experimental Life.* Princeton, NJ: Princeton University Press.

Sivado, Akos. 2015. 'The shape of *things* to come: reflections on the ontological turn in anthropology'. *Philosophy of the Social Sciences* 45 (1): 83–99.

Skafish, Peter. 2014. 'Introduction', in E. Viveiros de Castro, *Cannibal Metaphysics: For a Post-Structural Anthropology,* pp. 9–38. Translated and edited by Peter Skafish. Minneapolis, MN: Univocal Publishing.

Skirbekk, Gunnar. 2015. 'Bruno Latour's anthropology of the moderns: a reply to Maniglier'. *Radical Philosophy* 189: 45–47.

Smith, William Cantwell. 1979. *Faith and Belief.* Princeton, NJ: Princeton University Press.

Sneath, David, Martin Holbraad and Morten Axel Pedersen. 2009. 'Technologies of the imagination: an introduction'. *Ethnos* 74 (1): 5–30.

Spencer, Jonathan. 1989. 'Anthropology as a kind of writing', *Man* 24 (1): 145–64.

Sperber, Dan. 1985. *On Anthropological Knowledge.* Cambridge: Cambridge University Press.

Stacey, Judith. 1988. 'Can there be a feminst ethnography?' *Women's Studies International Forum* 11 (1): 21–27.

Stasch, Rupert. 2003. 'Separateness as a relation: the iconicity, univocality and creativity of Korowai mother-in-law avoidance'. *Journal of the Royal Anthropological Institute* 9: 317–37.

    2009. *Society of Others. Kinship and Mourning in a West Papuan Place.* Berkeley, CA: University of California Press.

Stengers, Isabelle. 2009. 'Experimenting with "What Is Philosophy?"', in C. B. Jensen and K. Rödje (eds.), *Deleuzian Intersections: Science, Technology, Anthropology,* pp. 39–56. New York: Berghahn.

    2010. *Cosmopolitics I.* Minneapolis, MN: University of Minnesota Press.

Stépanoff, Charles. 2009. 'Devouring perspectives: on cannibal shamans in Siberia'. *Inner Asia* 11: 283–307.

Stocking, George. 1984. 'Dr. Durkheim and Mr. Brown: comparative sociology at Cambridge in 1910' and 'Radcliffe-Brown and British Social Anthropology', in G. Stocking (ed.), *Functionalism Historicized. Essays on British Social Anthropology,* pp. 106–91. Madison, WI: University of Wisconsin Press.

    (ed.). 1986. *Malinowski, Rivers, Benedict and Others: Essays on British Social Anthropology.* History of Anthropology, Vol. 4. Madison: University of Wisconsin Press.

    1987. *Victorian Anthropology.* New York: Free Press, and London: Collier Macmillan.

    1995. *After Tylor: British Social Anthropology 1888–1951.* Madison, WI: *University of Wisconsin Press.*

Stoller Paul. 1995. *Spirit Possession, Power, and the Hauka in West Africa.* New York: Routledge.

Strathern, Marilyn. 1980. 'No nature, no culture: the Hagen case', in C. MacCormack and M. Strathern (eds.), *Nature, Culture and Gender,* pp.174–222. Cambridge: Cambridge University Press.

    1987a. 'An awkward relationship: the case of feminism in anthropology'. *Signs* 12 (2): 276–92.

    1987b. 'Out of context: the pervasive fictions of anthropology'. *Current Anthropology* 28 (3): 251–81.

Bibliography

1988. *The Gender of the Gift: Problems with Women and Problems with Society in Melanesia*. Berkeley, CA: University of California Press.

1990. 'Artefacts of history: events and the interpretation of images', in J. Siikala (ed.), *Culture and History in the Pacific*, pp. 25–44. Helsinki: Finnish Anthropological Society.

1992a. *After Nature: English Kinship in the Late Twentieth Century*. Cambridge: Cambridge University Press.

1992b. 'Parts and wholes: refiguring relationships in a postplural world', in A. Kuper (ed.), *Conceptualizing Society*. London: Routledge.

1995a. *The Relation: Issues in Complexity and Scale*. Cambridge: Prickly Pear Pamphlet no. 6.

1995b. 'Gender: division or comparison?', in N. Charles and F. Hughes-Freeland (eds.), *Practising Feminism: Identity, Difference, Power*, pp. 38–60. London: Routledge.

1996. 'Cutting the network'. *Journal of the Royal Anthropological Institute (N.S.)* 2 (3): 517–35.

1999. *Property, Substance and Effect. Anthropological Essays on Persons and Things*. London: The Athlone Press.

2004. *Partial Connections*. Updated edition. Oxford: Altamira Press.

2005. *Kinship, Law and the Unexpected: Relatives Are Always a Surprise*. Cambridge: Cambridge University Press.

2011. 'Binary license'. *Common Knowledge* 17 (1): 87–103.

2012. 'A comment on "the ontological turn" in Japanese anthropology'. *HAU: Journal of Ethnographic Theory* 2 (2): 402–5.

2014a. 'Reading relations backwards'. *Journal of the Royal Anthropological Institute (N.S.)* 20: 3–19.

2014b. 'Anthropological reasoning: some threads of thought'. *Hau: Journal of Ethnographic Theory* 4 (3): 23–37.

Street, Alice. 2014. *Biomedicine in an Unstable Place: Infrastructure and Personhood in a Papua New Guinean Hospital*. London and Durham, NC: Duke University Press.

Surel, Olivier. 2014. 'Let a thousand natures bloom: a polemical trope in the "ontological turn" of anthropology'. *Krisis: Journal for Contemporary Philosophy* 2: 14–29.

Suzuki, Wakana. 2015. 'The care of the cell: onomatopoeia and embodiment in a stem cell laboratory'. *NatureCulture* 3: 87–105.

Swancutt, Katherine. 2012. *Fortune and the Cursed: The Sliding Scale of Time in Mongolian Divination*. Oxford: Berghahn.

Swanson, Heather Anne, Nils Bubandt and Anna Tsing. 2015. 'Less than one but more than many: anthropocene as science fiction and scholarship-in-the-making'. *Environment and Society: Advances in Research* 6 (1): 149–66.

Tarde, Gabriel. 2012. *Monadology and Sociology*. Edited and translated by Theo Lorenc. Prahan: re.press.

Taussig, Michael. 1993. *Mimesis and Alterity. A Particular History of the Senses*. New York: Routledge.

Taylor, Christopher C. 2001. *Sacrifice as Terror: The Rwandan Genocide of 1994*. London: Bloomsbury Academic.

Taylor, Diana. 2014. *Michel Foucault: Key Concepts*. London: Routledge.

Thompson, Charis. 1996. 'Ontological choreography: agency through objectification in infertility clinics'. *Social Studies of Science* 26:575–610.

Thrift Nigel. 2007. *Non-representational Theory: Space, Politics, Affect*. London: Routledge.

Tresch, John. 2013. 'Another turn after ANT: an interview with Bruno Latour'. *Social Studies of Science* 43 (2): 302–13.

Tsing, Anna L. 2005. *Friction. An Ethnography of Global Connection*. Princeton, NJ: Princeton University Press.

2013. 'More-than-human sociality: a call for critical description', in K. Hastrup (ed.), *Anthropology and Nature*, pp. 27–42. London: Routledge.

2014. 'Strathern beyond the human: testimony of a spore'. *Theory, Culture & Society* 31 (2–3): 321–44.

2015. *The Mushroom at the End of the World: On the Possibility of Life in Capitalist Ruins*. Princeton, NJ: Princeton University Press.

Turner, Terence. 2009. 'The crisis of late structuralism, perspectivism and animism: rethinking culture, nature, spirit, and bodiliness'. *Tipití: Journal of the Society for the Anthropology of Lowland South America* 7 (1): 3–42.

Tyler, Stephen. 1987. *The Unspeakable: Discourse, Dialogue and Rhetoric in the Postmodern World*. Madison, WI: University of Wisconsin Press.

Tylor, Edward B. 1920. *Primitive Culture*. Volume 1. New York: J. P. Putnam's Sons.

Uexküll, Jacob von. 1934. 'A stroll through the words and animals and men', in C. H. Schiller and K. S. Lashley (eds.), *Instinctive Behavior: The Development of a Modern Concept*, pp. 5–80. New York: International Universities Press.

Uzendosky, Michael. 2012. 'Beyond orality: textuality, territoriality, and ontology among Amazonian people'. *Hau: Journal of Ethnographic Theory* 2 (1): 55–80.

Valeri, Valerio. 1995. 'Miti cosmogonici e ordine'. *Parole Chiave* 7/8: 93–110.

Vangkilde, Kasper T. 2015. 'Possessed by the Zeitgeist: inspiration and prophecy in the business of fashion'. *Journal of Business Anthropology* 4 (2): 178–200.

Venkatesan, Soumyha, Michael Carrithers, Karen Sykes, Matei Candea and Martin Holbraad. 2010. '"Ontology is just another word for culture": motion tabled at the 2008 meeting of the Group for Debates in Anthropological Theory'. *Critique of Anthropology* 30: 152–200.

Bibliography

Venkatesan, S., M. C. C. B. Jensen, J. Leach, M. A. Pedersen, and G. Evans. 2012. 'The task of anthropology is to invent relations: 2011 meeting of the Group for Debates in Anthropological Theory'. *Critique of Anthropology* 32 (1): 43–86.

Verran, Helen. 2001. *Science and an African Logic*. Chicago: University of Chicago Press.

——— 2014. 'Anthropology as ontology is comparison as ontology'. Fieldsights – Theorizing the Contemporary. *Cultural Anthropology* website, http://culanth .org/fieldsights/468-anthropology-as-ontology-is-comparison-as-ontology (accessed 13 January 2014).

Vigh, Henrik E. 2006. *Navigating Terrains of War: Youth and Soldiering in Guinea-Bissau*. Oxford: Berghahn.

Vigh, Henrik E. and David Brehm Sausdal. 2014. 'From essence back to existence: anthropology beyond the ontological turn'. *Anthropological Theory* 14 (1): 49–73.

Vilaça, Aparecida. 2005. 'Chronically unstable bodies: reflections on Amazonian corporalities'. *Journal of Royal Anthropological Institute (N.S.)*, 11.3 (September 2005): 445–64.

——— 2011. 'Dividuality in Amazonia: God, the Devil, and the constitution of personhood in Wari' Christianity'. *Journal of the Royal Anthropological Institute (N.S.)* 17: 243–62.

Viveiros de Castro, Eduardo. 1992. *From the Enemy's Point of View: Humanity and Divinity in an Amazon Society*. Chicago: University of Chicago Press.

——— (ed.). 1995. *Pensando o parentesco ameríndio: estudos amerindios*. Rio de Janeiro: UFRJ.

——— 1998. 'Cosmological deixis and Amerindian perspectivism'. *Journal of Royal Anthropological Institute* 4 (3): 469–88.

——— 2003. *AND*. Manchester: Manchester Papers in Social Anthropology 7.

——— 2004. 'Perspectival anthropology and the method of controlled equivocation'. *Tipití: Journal of the Society for the Anthropology of Lowland South America* 2 (1): 3–22.

——— 2007. 'The Crystal Forest: notes on the ontology of Amazonian spirits'. *Inner Asia* 9 (2): 153–72.

——— 2009. 'Intensive filiation and demonic alliance', in C. B. Jensen and K. Rödje (eds.), *Deleuzian Intersections in Science, Technology and Anthropology*, pp. 219–55. Oxford: Berghahn Books.

——— 2010. 'Claude Lévi-Strauss: fundador del postestructuralismo', in M. E. Olavarria, C. Bonfiglioli, S. Millan (eds.), *Lévi-Strauss: un siglo de reflexion*, pp. 17–42. Districto Fedreral de Mexico: Juan Pablos Editor.

——— 2011a. 'Zeno and the art of anthropology: of lies, beliefs, paradoxes, and other truths'. *Common Knowledge* 17 (1): 128–45.

2011b. *The Inconstancy of the Indian Soul: The Encounter of Catholics and Cannibals in 16th Century Brazil*. Chicago: Prickly Paradigm Press.

2012. *Cosmological Perspectivism in Amazonia and Elsewhere*. Masterclass Series 1. Manchester: HAU Network of Ethnographic Theory.

2013. 'The relative native'. *Hau: Journal of Ethnographic Theory* 3 (3): 473–502.

2014. *Cannibal Metaphysics: For a Post-Structural Anthropology*. Translated and edited by Peter Skafish. Minneapolis, MN: Univocal Publishing.

2015. Who is afraid of the ontological wolf? Some comments on an ongoing anthropological debate. *The Cambridge Journal of Anthropology* 33 (1): 2–17.

Viveiros de Castro, Eduardo and Carlos Fausto. 1993. 'La puissance et l'acte: la parenté dans les basses terres de l'Amérique du Sud'. *L'Homme* XXXIII (2–4): 141–70.

Viveiros de Castro, Eduardo and Marcio Goldman. 2009. 'Slow motions: comments on a few texts by Marilyn Strathern'. *Cambridge Anthropology* 28 (3): 23–42.

Wagner, Roy. 1972. *Habu: The Innovation of Meaning in Daribi Religion*. Chicago: University of Chicago Press.

1977a. 'Scientific and indigenous Papuan conceptualizations of the innate: a semiotic critique of the ecological perspective', in T. P. Bayliss-Smith and R. G. A. Feachern (eds.), *Subsistence and Survival: Rural Ecology in the Pacific*, pp. 385–410. London: Academic Press.

1977b. 'Analogic kinship: a Daribi example'. *American Ethnologist* 4 (4): 623–42.

1978. *Lethal Speech: Daribi Myth As Symbolic Obviation*. Ithaca: Cornell University Press.

1981. *The Invention of Culture – revised and expanded edition*. Chicago: University of Chicago Press.

1984. 'Ritual as communication: order, meaning, and secrecy in Melanesian initiation rites'. *Annual Review of Anthropology* 13: 143–55.

1986a. *Symbols That Stand for Themselves*. Chicago: University of Chicago Press.

1986b. *Asiwinarong: Ethos, Image, and Social Power Among the Usen Barok of New Ireland*. Princeton, NJ: Princeton University Press.

1987. 'Figure-ground reversal among the Barok', in L. Lincoln (ed.), *Assemblage of Spirits: Idea and Image in New Ireland*, pp. 56–62. New York: George Braziller.

1991. 'The fractal person', in M. Godelier and M. Strathern (ed.), *Big Men and Great Men: Personifications of Power in Melanesia*, pp. 159–74. Cambridge: Cambridge University Press.

2000. 'Condensed mapping: myth and the folding of space / space and the folding of myth', in A. Ramsey and J. F. Weiner (eds.), *Emplaced Myth: Space, Narrative and Knowledge in Aboriginal Australia and Papua New Guinea*, pp. 71–8. Honolulu: University of Hawai'i Press.

Bibliography

2001. *An Anthropology of the Subject: Holographic Worldview in New Guinea and Its Meaning and Significance for the World of Anthropology.* Berkeley, CA: University of California Press.

2005. 'Afterword: order is what happens when chaos loses its temper', in M. S. Mosko and F. Damon (eds.), *On the Order of Chaos: Social Anthropology and the Science of Chaos*, pp. 206–47. New York and London: Berghahn Books.

2010. *Coyote Anthropology.* Lincoln and London: University of Nebraska Press.

2012a. 'Facts force you to believe *in* them; perspectives encourage you to believe *out* of them: an introduction to Viveiros de Castro's magisterial essay', in E. Viveiros de Castro *Cosmological Perspectivism in Amazonia and Elsewhere.* Master class Series 1. Manchester: HAU Network of Ethnographic Theory.

2012b. '"Luck in the double focus": ritualized hospitality in Melanesia'. *Journal of the Royal Anthropological Institute* 18: 161–74.

Walford, Antonia. 2012. 'Data moves: taking Amazonian climate science seriously'. *Cambridge Anthropology* 30 (2): 101–17.

2015. 'Double standards: examples and exceptions in scientific meteorological practice in Brazil'. *Journal of the Royal Anthropological Institute (N.S.)* 21 (S1): 64–77.

Wastell, Sari. 2001. 'Presuming scale, making diversity'. *Critique of Anthropology* 21 (2): 185–210.

2007. 'The "legal" thing in Swaziland: *res judicata* and divine kinship', in A. Henare, M. Holbraad and S. Wastell (eds.), *Thinking Through Things. Theorizing Artefacts Ethnographically*, pp. 68–92. London: Routledge.

Watson, Matthew C. 2014. 'Derrida, Stengers, Latour and subalternist cosmopolitics'. *Theory, Culture & Society* 31 (1): 75–98.

Weber, Max. 1958. *The Protestant Ethic and the Spirit of Capitalism.* New York: Scribner

1968. *Economy and Society.* Edited by Guenther Roth and Claus Wittich. Berkeley, CA: University of California Press.

1992. *The Protestant Ethic and the Spirit of Capitalism.* London: Routledge.

Webmoore, Tim and Christopher Witmore. 2008. 'Things are us! A commentary on human/things relations under the banner of a "social" archaeology'. *Norwegian Archaeology Review*, 41 (1), 53–70.

Weiner, Annette B. 1992. *Inalienable Possessions: The Paradox of Keeping-While Giving.* Berkeley, CA: University of California Press.

Weiner, James F. 1988. *The Heart of the Pear Shell: The Mythological Dimension of Foi Sociality.* Berkeley, CA and London: University of California Press.

1995. *The Lost Drum: The Myth of Sexuality in Papua New Guinea and Beyond.* Madison, WI: The University of Wisconsin Press.

Whatmore, Sarah J. 2013. 'Earthly powers and affective environments: an ontological politics of flood risk'. *Theory, Culture & Society* 30 (7/8): 33–50.

Willerslev, Rane. 2004. 'Not animal not not-animal: hunting, imitation and empathetic knowledge among the Siberian Yukaghirs'. *Journal of the Royal Anthropological Institute* 10: 629–52.

2007. *Soul Hunters. Hunting, Animism, and Personhood among the Siberian Yukaghirs*. Berkeley, CA: University of California Press.

2011. 'Frazer strikes back from the armchair: a new search for the animist soul'. *Journal of the Royal Anthropological Institute* 17 (3): 504–26.

Willerslev, Rane and Morten A. Pedersen. 2010. 'Proportional holism: joking the cosmos into the right shape in Northern Asia', in T. Otto and N. Bubandt (eds.), *Experiments with Holism*, pp. 162–78. London: Blackwell.

Wilson, Bryan R. (ed.). 1974. *Rationality*. Oxford: Basil Blackwell.

Wimsatt, William C. 2007. *Re-engineering Philosophy for Limited Beings: Piecewise Approximations to Reality*. Cambridge, MA: Harvard University Press.

Winch, Peter. 1967. *The Idea of Social Science and its Relation to Philosophy*. London and New York: Routledge and Keegan Paul.

Winthereik, Brit R. and Helen Verran. 2012. 'Stories as generalizations that intervene'. *Social Science Studies* 25 (1): 37–51.

Witmore, Christopher. 2009. 'The realities of the past', in B. Fortenberry & L. McAtackney (eds.), *Modern Materials: Proceedings from the Contemporary and Historical Archaeology in Theory Conference*, pp. 25–36. Oxford: British Archaeological Reports.

In press. 'The realities of the past: archaeology, object-orientations, pragmatology', in B. R. Fortenberry and L. McAtackney (eds.), *Modern Materials: Proceedings from the Contemporary and Historical Archaeology in Theory Conference 2009*. Oxford: Archaeopress.

Woolgar, Steve and Javier Lezaun. 2013. 'The wrong bin bag: a turn to ontology in science and technology studies?' *Social Studies of Science* 43 (3): 321–40.

Zhan Mei. 2009. *Other-Worldly: Making Chinese Medicine through Transnational Frames*. Durham, NC: Duke University Press.

Index

Index

Index